Gas Turbine Technology

Gas Turbine Technology

Edited by **Eugene Bradley**

New York

Published by NY Research Press,
23 West, 55th Street, Suite 816,
New York, NY 10019, USA
www.nyresearchpress.com

Gas Turbine Technology
Edited by Eugene Bradley

© 2015 NY Research Press

International Standard Book Number: 978-1-63238-213-9 (Hardback)

Printed in the United States of America.

Contents

Preface

Over the recent decade, advancements and applications have progressed exponentially. This has led to the increased interest in this field and projects are being conducted to enhance knowledge. The main objective of this book is to present some of the critical challenges and provide insights into possible solutions. This book will answer the varied questions that arise in the field and also provide an increased scope for furthering studies.

Gas turbine engines will be the dominant essential technology in the next 20-year energy scenarios, either in stand-alone procedures or in combination with other energy generation apparatus. This book gives a comprehensive summary of gas turbine technology and describes some of the key developments that feature the gas turbine technology in various applications, like marine and aircraft propulsion, and industrial and stationary power generation. Thus, this book targets design, maintenance, analyst, and material engineers. Also, it will be highly beneficial to manufacturers, researchers and scientists due to the timely and correct knowledge presented in this book.

I hope that this book, with its visionary approach, will be a valuable addition and will promote interest among readers. Each of the authors has provided their extraordinary competence in their specific fields by providing different perspectives as they come from diverse nations and regions. I thank them for their contributions.

Editor

Part 1

Aero and Marine Gas Turbines

Future Aero Engine Designs: An Evolving Vision

Konstantinos G. Kyprianidis
Chalmers University of Technology
Sweden

1. Introduction

Public awareness and political concern over the environmental impact of civil aviation growth has improved substantially during the past 30 years. As the environmental awareness increases, so does the effort associated with addressing NO_x and CO_2 emissions by all the parties involved. In the Vision 2020 report made by the Advisory Council for Aeronautical Research in Europe (2001), goals are set to reduce noise and emissions produced by the ever increasing global air traffic. Emissions legislation, set by the International Civil Aviation Organisation (ICAO) and it's Committee on Aviation Environmental Protection (CAEP), is becoming ever more stringent, creating a strong driver for investigating novel aero engine designs that produce less CO_2 and NO_x emissions.

On the other hand, airline companies need to continuously reduce their operating costs in order to increase, or at least maintain, their profitability. This introduces an additional design challenge as new aero engine designs need to be conceived for reduced environmental impact as well as direct operating costs. Decision making on optimal engine cycle selection needs to consider mission fuel burn, direct operating costs, engine and airframe noise, emissions and global warming impact.

CO_2 emissions are directly proportional to fuel burn, and therefore any effort to reduce them needs to focus on improving fuel burn, by reducing engine Specific Fuel Consumption (SFC), weight and size. Reducing engine weight results in a lower aircraft maximum take-off weight, which in turn leads to reduced thrust requirements for a given aircraft lift to drag ratio. Reducing engine size – predominantly engine nacelle diameter and length – reduces nacelle drag and therefore also leads to reduced thrust requirements. For a given engine SFC, a reduction in thrust requirements essentially results in lower fuel burn. Lower engine SFC can be achieved by improving propulsive efficiency and thermal efficiency – either by reducing component losses or by improving the thermodynamic cycle.

Improvements in propulsive efficiency – and hence engine SFC at a given thermal efficiency – can be achieved by designing an engine at a lower specific thrust (i.e. net thrust divided by fan inlet mass flow). This results in a larger fan diameter, at a given thrust, and therefore in increased engine weight, which can partially, or even fully, negate any SFC benefits. Propulsive efficiency improvements at a constant weight are directly dependent on weight reduction technologies such as light weight fan designs and new shaft materials. Increasing engine bypass ratio aggravates the speed mismatch between the fan and the low pressure turbine. Introduction of a gearbox can relieve this issue by permitting the design of these two components at their optimal speeds, and can hence reduce engine weight, as well as improve component efficiency. The first research question therefore rises:

How low can we really go on specific thrust?

Improvements in thermal efficiency – and hence engine SFC at a given propulsive efficiency – can be achieved for conventional cores mainly by increasing engine Overall Pressure Ratio (OPR). At a given OPR there is an optimal level of combustor outlet temperature T_4 for thermal efficiency. However, at a fixed specific thrust and engine thrust, an increase in T_4 can result in a smaller core and therefore a higher engine bypass ratio; in some cases, a potential reduction in engine weight can more than compensate for a non-optimal thermal efficiency. Increasing OPR further than current engine designs is hindered by limitations in high pressure compressor delivery temperature at take-off. Increasing T_4 is limited by maximum permissable high pressure turbine rotor metal temperatures at take-off and top of climb. Increasing turbine cooling flows for this purpose is also fairly limited as a strategy; cooling flows essentially represent losses in the thermodynamic cycle, and increasing them eventually leads to severe thermal efficiency deficits (Horlock et al., 2001; Wilcock et al., 2005). Designing a combustor at very low air to fuel ratio levels is also limited by the need for adequate combustor liner film-cooling air as well as maintaining an acceptable temperature traverse quality (Lefebvre, 1999). The second research question therefore rises:

How high can we really go on OPR and T_4?

Aggressive turbofan designs that reduce CO_2 emissions – such as increased OPR and T_4 designs – can increase the production of NO_x emissions due to higher flame temperatures. The third research question therefore rises:

What is the trade-off between low CO_2 and NO_x?

The research work presented in this chapter will focus on identifying several novel engine cycles and technologies - currently under research - that can address the three research questions raised. These concepts will be evaluated based on their potential to reduce CO_2 and NO_x emissions for engine designs entering service between 2020 and 2025. Design constraints, material technology, customer requirements, noise and emissions legislation, technology risk and economic considerations and their effect on optimal concept selection will also be discussed in detail.

2. An evolving vision

Numerous feasibility studies have been published over the years focusing on future engine and aircraft designs that can reduce fuel consumption; a brief review of some of these publications will be carried out here.

One of the earliest discussions on the subject of improving engine fuel efficiency is provided by Gray & Witherspoon (1976), looking at conventional and heat exchanged cores, as well as non-steady flow combustion processes and open rotor configurations. A similar study focusing on geared and open rotor arrangements as well as heat exchanged cycles is presented by Hirschkron & Neitzel (1976).

An interesting discussion on how specific thrust levels were expected to evolve in the mid-70's based on the economic and technological projections of that time period is given by Jackson

(1976); the author has also provided an update to that discussion based on current economical and technological projections (Jackson, 2009). Wilde (1978), Young (1979), and Pope (1979) provide a good reference on how the future for civil turbofan engines for medium and long range applications was envisaged in the late 70's. Some early discussions on future trends in commercial aviation from the aircraft manufacturer's perspective can be found in Swihart (1970) and Bates & Morris (1983), while Watts (1978) provides an airliner's view of the future. A review on the several technical and economic obstacles that were identified in the late 80's with respect to the realization of the Ultra-High Bypass Ratio (UHBR) turbofan concept is provided by Borradaile (1988) and by Zimbrick & Colehour (1988). Peacock & Sadler (1992) give an update on the subject, focusing further on engine design constraints and the technology advancements required for producing a competitive UHBR configuration. Potential year 2020 scenarios are explored by Birch (2000) while an overview of current aero engine technology and some insight on the future of aircraft propulsion is given by Ruffles (2000). Sieber (1991) and Schimming (2003) provide an excellent discussion on counter-rotating fan designs. Finally, for a review on the development of civil propulsion from the early 50's to recent years the interested reader is referred to Saravanamuttoo (2002). The focus of the next section will be given on recent European research initiatives on enabling technologies relevant to the three research questions that have been set.

3. Enabling technologies and recent research

3.1 Propulsor technologies

Within the EU Framework Program 6 research project VITAL (enVIronmenTALly friendly aero engines, 2009) a number of low pressure system component technologies have been investigated (Korsia, 2009; Korsia & Guy, 2007). The emerging progress will allow the design of new powerplants capable of providing a step reduction in fuel consumption and generated noise.

The VITAL project concentrated on new technologies for the low pressure system of the engine, which enable the development of low noise and low weight fan architectures for UHBR engines. To achieve these objectives, the VITAL project has investigated three different low pressure configurations, leading to low noise and high efficiency power plants. The three configurations are the DDTF (Direct Drive TurboFan) supported by Rolls-Royce, the GTF (Geared TurboFan) by MTU and the CRTF (Counter-Rotating TurboFan) by Snecma.

The DDTF architecture offers a re-optimised trade-off between fan and turbine requirements considering the low weight technologies introduced by the VITAL programme. The GTF combines a fan with a reduction gear train, to allow different rotating speeds for the fan on one hand, and the booster and turbine on the other. The CRTF offers a configuration with two fans turning in opposite directions, allowing for lower rotational speeds, since the two fan rotors split the loads involved.

The technologies being built into the VITAL engines include (Korsia, 2009; Korsia & Guy, 2007):

- New fan concepts with the emphasis on two types: counter-rotating and lightweight fans.

- New booster technologies for different operational requirements; low and high speed, associated aerodynamic technologies, new lightweight materials and associated coating and noise reduction design.

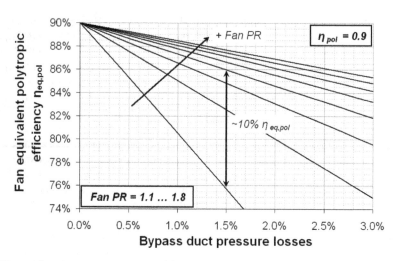

Fig. 1. Effect of fan tip pressure ratio and bypass duct pressure losses on fan equivalent polytropic efficiency

- Polymer composites and corresponding structural design and manufacturing techniques are studied in parallel with advances in metallic materials and manufacturing processes.
- Shaft torque density capabilities through the development of metal matrix composites and multi metallic shafts.
- Low pressure turbine weight savings through ultra high lift airfoil design, ultra high stage loading, lightweight materials and design solutions.
- Technologies for light weight and low drag installation of high bypass ratio engines related to nozzle, nacelle and thrust reverser.

The open rotor engine concept, for high subsonic flight speeds, has also risen as a candidate for improving fuel consumption on several occasions since the advent of the first high bypass ratio turbofan engine. Such engine configurations, often refereed to as propfans in the literature, are direct competitors to ultra high bypass ratio turbofan engines. Their are located at the ultra-low specific thrust region of the design space, where propulsive efficiency benefits for turbofans are negated by very low transfer efficiencies. As illustrated in Fig. 1, this is due to the dominant effect on transfer efficiency that bypass duct pressure losses have when looking at low fan tip pressure ratio engine designs, i.e. low specific thrust. Open rotor engines do not suffer from bypass duct pressure losses and can therefore achieve a very high propulsive efficiency at a good level of transfer efficiency. Compared to turbofans, propfans also benefit from reduced nacelle drag and weight penalties.

Several open rotor programs took place during the 80's, resulting in engine demonstrators and flight tests. The purpose of these projects was to develop propfan concepts that could fly efficiently at speeds comparable to high bypass ratio turbofans, i.e. close to Mach 0.8. General Electric proposed the UDF (UnDucted Fan), a pusher configuration with counter-rotating propellers driven by a counter-rotating low pressure turbine (GE36 Design and Systems Engineering, 1987). The 578-DX, a pusher configuration with counter-rotating

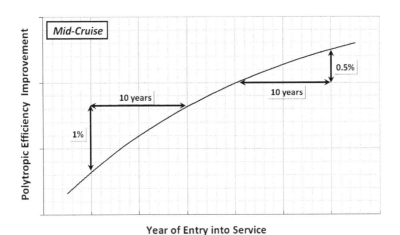

Fig. 2. Compressor efficiency improvement with year of entry into service.

propellers driven by a more conventional low pressure turbine through a differential planetary differential gearbox, was the result of a joint effort by Pratt & Whitney, Hamilton Standard, and Allison.

Both projects were eventually put on hold towards the end of the decade as fuel prices fell significantly. Nevertheless, the open rotor concept has now resurfaced within the EU Framework Program 7 research project DREAM (valiDation of Radical Engine Architecture systeMs, 2011) and the Clean Sky Joint Technology Initiative (2011). Within DREAM, the feasibility of two different open rotor architectures is evaluated including noise. Within Clean Sky, research work is being carried out by some of Europe's largest aero engine manufacturers, such as Rolls-Royce and Snecma, focused on designing, building and testing an open rotor demonstrator.

3.2 Core technologies

Improving core component efficiencies (including reducing losses in the cycle such as duct pressure losses) is one way of improving the engine thermal efficiency. Nevertheless, modern CFD-assisted designs are already quite aggressive and limited benefit may be envisaged by such future advancements (Kurzke, 2003); the increasing effort required to improve an already very good axial compressor design is illustrated in Fig.2.

Within the EU Framework Program 6 research project NEWAC (NEW Aero engine Core concepts, 2011) a number of advanced core component technologies have been investigated that include (Rolt & Kyprianidis, 2010; Wilfert et al., 2007):

- Improved high pressure compressor aero design and blade tip rub management.
- Flow control technologies including aspirated compression systems.
- Active control of surge and tip clearance in compressors.
- Active control of a cooled cooling air system.

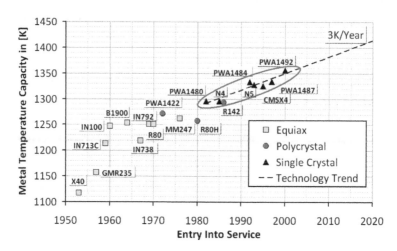

Fig. 3. Evolution of turbine material capability and future trend.

As mentioned earlier another way of improving engine thermal efficiency is to raise the cycle OPR. For conventional cores, increasing OPR and T_4 depends primarily on future advancements in material and cooling technology. The evolution of turbine material capability over a period of 50 years is illustrated in Fig. 3. As can be observed, only mild improvements have been achieved so far and this seems to be a continuing trend; the potential introduction of ceramics would form a major improvement in the field, but substantially more research is still required before realising this. Despite the low improvement rate in turbine material technology (roughly 3 [K/year]) aero engine designs have seen substantial increases in T_4 over the last 60 years (roughly 10 [K/year]); this is illustrated in Fig. 4 for engines designed for long-haul applications. The main reason behind these improvements in T_4 has been the introduction of cooling and Thermal Barrier Coatings (TBC) in turbine designs; the interested reader is referred to Downs & Kenneth (2009) for a good overview of the evolution of turbine cooling systems design.

It is perhaps debatable whether an improvement rate of 10 [K/year] in T_4 can be maintained in the future, and for that reason the design focus for more aggressive thermal efficiency improvements could very well be redirected to the introduction of heat-exchanged cores and advanced compressor technologies for future turbofan designs. In that respect, some of the technologies researched under the NEWAC project can be perceived as intermediate enabling steps for realising new engine core concepts that could improve the core thermal efficiency. These new core concepts comprise of:

- Ultra-high OPR core with intercooling.
- Medium OPR intercooled recuperated core.
- High OPR flow controlled core.
- High OPR active core including active cooling air cooling.

When considering intercooling for an aero engine design, a common textbook misconception is that the thermal efficiency of an intercooled core will always be lower than a conventional

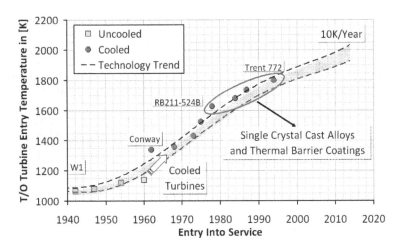

Fig. 4. Evolution of turbine entry temperature and future trend.

core's for a fixed OPR and specific thrust (Saravanamuttoo et al., 2001). The argument behind this is that the heat removed by the intercooler will largely need to be reintroduced in the combustor by burning more fuel, while the reduction in compression work and increase in bypass stream thrust (due to the heat rejection) will only partially compensate for the loss in cycle efficiency, at a fixed specific thrust and T_4. Adding the expected intercooler pressure losses in the cycle calculations would further worsen the SFC deficit and make the increase in specific thrust less marked.

However, cycle calculations based on half-ideal gas properties and no dissociation (i.e. isobaric heat capacity dependent on temperature), presented by Walsh & Fletcher (1998), give a slightly different picture on intercooling. For a given T_4, the optimal OPR for an intercooled core will be much higher than that for a conventional core. Comparing the two concepts at their optimal OPR levels, for a given technology level, can make the intercooled core more attractive with respect to thermal efficiency and not just specific thrust. Canière et al. (2006) and da Cunha Alves et al. (2001) also reached the same conclusion about the thermal efficiency of the intercooled cycle while studying this concept for gas turbines used in power generation. Papadopoulos & Pilidis (2000) worked on the introduction of intercooling, by means of heat pipes, in an aero engine design for long haul applications. Xu et al. (2007) performed a mission optimization to assess the potential of a tubular intercooler. Recent work by Xu & Grönstedt (2010) presents a refined tubular configuration estimating a potential block fuel benefit of 3.4%. The work addresses the limitation that short high pressure compressor blade lengths and related low compression efficiencies may impose on engines designed for short range missions, and suggest a novel gas path layout as a remedy to this constraint. A design study of a high OPR intercooled aero engine is described in Rolt & Baker (2009), while details on the aerodynamic challenges in designing a duct system to transfer the core air into and out of the intercooler are presented by Walker et al. (2009).

The introduction of recuperation in an aero engine, for high thermal efficiency at low OPR, has also been the focus of different researchers. Lundbladh & Sjunnesson (2003) performed a feasibility study for InterCooled (IC) and Intercooled Recuperated Aero engines (IRA) that consider cycle benefits, weights and direct operating costs. Boggia & Rud (2005) provide an extended discussion on the thermodynamic cycle and the technological innovations necessary for realizing the intercooled recuperated core concept. Various aspects of the thermo-mechanical design of a compact heat exchanger have been presented by Pellischek & Kumpf (1991) and Schoenenborn et al. (2006). For a comprehensive review on the development activities for recuperated aero engines since the late 60's the interested reader can refer to McDonald et al. (2008a;b;c).

Finally, three different types of lean-burn combustor technology were also researched within NEWAC with the objective of reducing emissions of oxides of nitrogen (NO_x):

- Lean Direct Injection (LDI) combustor for high and ultra-high OPR cores.
- Partial Evaporation and Rapid Mixing (PERM) combustor for high OPR cores.
- Lean Premixed Pre-vaporized (LPP) combustor for medium OPR cores.

4. Design space exploration

4.1 Methodology, design feasibility and constraints

To effectively explore the design space a tool is required that can consider the main disciplines typically encountered in conceptual design. The prediction of engine performance, aircraft design and performance, direct operating costs, and emissions for the concepts analysed in this study was made using the EVA code (Kyprianidis et al., 2008). Another code, WeiCo, was also used for carrying out mechanical and aerodynamic design in order to derive engine component weight and dimensions. The two tools have been integrated together within an optimiser environment as illustrated in Fig. 5, based on lessons learned from the development of the TERA2020 tool (Kyprianidis et al., 2011). This integration allows for multi-objective optimization, design studies, parametric studies, and sensitivity analysis. In order to speed up the execution of individual engine designs, the conceptual design tool attempts to minimize internal iterations in the calculation sequence through the use of an explicit algorithm, as described in detail by Kyprianidis (2010).

Aero-engine designs are subject to a large number of constraints and these need to be considered during conceptual design. Constraints can be applied within the optimiser environment at the end of the calculation sequence i.e., after the last design module has been executed. During a numerical optimisation, the optimiser will select a new set of input design parameters for every iteration and the resulting combination of aircraft and engine will be assessed. Using user specified objective functions the optimiser will home in on the best engine designs, determining the acceptability/feasibility of each design through the constraints set by the user. Infeasible designs will be ruled out, while non-optimum design values will result in engine designs with non-optimum values for the objective function selected. The optimiser will therefore avoid regions in the design pool that result in infeasible or non-optimum engine designs.

Design constraints set by the user include among others:

- Take-off HPC delivery temperature and other important performance parameters.
- FAR (Federal Aviation Regulations) take-off field length for all engines operating and balanced field length for one engine inoperative conditions.
- Time to height.

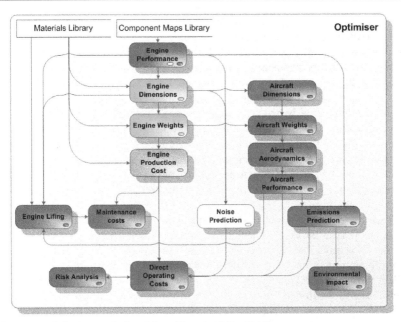

Fig. 5. Conceptual design tool algorithm (Kyprianidis, 2010).

- LTO (Landing and Take-Off) cycle D_pNO_x/F_{oo} vs. ICAO certification limits and CAEP medium and long term goals.

- Cumulative EPNL vs. ICAO certification limits.

- Engine time between overhaul.

Where component design is concerned, for a conventional core the High Pressure Compressor (HPC) delivery temperature, and hence the engine OPR, is typically constrained by the mechanical properties of the HPC disc or HPC rear drive cone or High Pressure Turbine (HPT) disc material (Rolt & Baker, 2009). For an intercooled core, the OPR value is no longer constrained by a maximum allowable HPC delivery temperature. Nevertheless, the intercooling process increases the air density in the gas path and as a result the compressor blades tend to become smaller. Losses from tip clearances become increasingly important and a minimum compressor blade height limitation needs to be applied to maintain state of the art compressor efficiency. Core architecture selections for the conventional core set an upper limit to the HPC design pressure ratio that can achieved when driven by a single-stage HPT. A transonic single-stage HPT design can allow for relatively higher HPC pressure ratios at the expense of a lower polytropic efficiency. A two-stage HPT can offer high HPC pressure ratios at a high polytropic efficiency but a trade-off arises with respect to the need for more cooling air and increased engine length associated with the introduction of a second row of vanes and blades. With respect to the intercooled core, the minimum design pressure ratio for the Intermediate Pressure Compressor (IPC) can in some cases be limited by icing considerations during the descent flight phase. The maximum area variation that may be achieved by the variable area auxiliary nozzle is also constrained by mechanical (and aerodynamic) considerations.

As discussed earlier, designing a combustor at very low air to fuel ratio levels is also limited by the need for adequate combustor liner film-cooling air as well as maintaining an acceptable temperature traverse quality (Lefebvre, 1999); this sets an upper bound on combustor outlet temperature. Furthermore, a maximum permissible mean metal temperature needs to be set to consider turbine blade material limitations. A lower bound on engine time between overhaul also needs to be set to limit the frequency of workshop visits. For short range applications the minimum engine time between overhaul was set to 18000 [hr] while for long range applications to 23000 [hr]. This reflects the fact that designs for short range applications are typically operated at high power conditions for a significantly larger part of their operational life. Significantly lower levels of maximum combustor outlet temperature and turbine blade mean metal temperature had to be selected, compared to what could be selected for engine designs for long range applications that are often operated at derated thrust levels and spend most of their life at cruise.

A rubberised aircraft wing model was used in these studies to capture 'snowball effects' with respect to maximum take-off weight variation, rather than using fixed engine thrust requirements. The aircraft drag polar and weight breakdown were predicted at component level from the aircraft geometry and high lift device settings for the take-off and approach phases. Fuel burned was calculated for the entire flight mission including reserves assuming ISA conditions, as illustrated in Fig. 6. Cruise is performed at the optimum altitude for specific range (fixed cruise Mach number) using a step-up cruise procedure as the aircraft gets lighter. A comprehensive take-off field length calculation is performed for all engines operating and one engine inoperative conditions up to 1500 [ft].

Two baseline aircraft models have been used herein; one model for long range applications and one for short range. The former model is largely based on public domain information available for the Airbus A330-200 while the latter model is based on the Airbus A320-200. The short range aircraft was designed to carry 150 [pax] for a distance of 3000 [nmi] and a typical business case of 500 [nmi]; for long range applications it was designed for 253 [pax], 6750 [nmi] and 3000 [nm], respectively. For the step-up cruise procedure, a minimum residual rate of climb of 300 [ft/min] was set as a constraint for flying at the cruise altitude for maximum specific range.

The maximum values for FAR take-off field length and time to height were set for a load factor of 1 and no cargo. The choice of both is based on customer operational requirements as the aircraft needs to be able to: (i) take-off from a large number of airports around the world and (ii) climb to the initial cruise altitude sufficiently fast to ease operations with local air traffic control (and hence reduce waiting time on the ground). A cumulative distribution of the world's major runway lengths, based on data from Jenkinson et al. (1999), is illustrated in Fig. 7. For short range applications fairly stringent constraints are typically set for the maximum take-off distance and time to height; in this study these were set to 2.0 [km] and 25 [min], respectively. For long range applications a maximum take-off distance of 2.5 [km] was set instead. Stringent constraints result in bigger engines but allow for greater flexibility for engine derating at a smaller block fuel cost.

The choice of load factor and cargo is considered sensible but it does not necessarily constitute a typical airline practice. Validating absolute block fuel predictions with public domain airline data is not a trivial task as different airlines will follow different operational practices. For example for the long range aircraft model, the business case prediction is 10% lower than the published annually-averaged value, given in [lt/(km*pax)], by SwissAir for 2009 for the

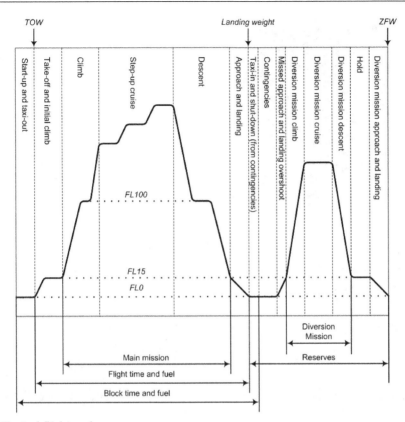

Fig. 6. Typical flight cycle.

Airbus A330-200 (Swiss International Air Lines, 2009). This does not necessarily mean that the model's business case is not a realistic one; nor that it wouldn't fit well with operational practices followed by other airlines. Furthermore, regional Air Traffic Management (ATM) practices can skew available block fuel data, while global ATM regulations may very well change significantly by 2020. It should be noted that fuel planning within the model respects the requirements defined for international flights by Federal Aviation Administration (n.d.) and Joint Aviation Authorities (2008).

Where conceptual design is concerned, exchange rates are perhaps a better type of parameter for evaluating the accuracy of a rubberised wing model, rather than just simply comparing absolute values. Block fuel exchange rates produced with the rubberised wing baseline aircraft models are presented in Table 1 for the business case of the long and short range models and are considered reasonable numbers.

During a block fuel optimization all engine aircraft combinations which do not fulfil the take-off and time to height criteria set will be discarded as infeasible. Due to the underlying physics, this will naturally lead to an optimal engine and aircraft combination for the defined objective function. All large engines will produce heavier aircraft with more drag and thus higher block fuel weight. Engines which are too small will not deliver enough thrust to satisfy the take-off and time to height criteria set.

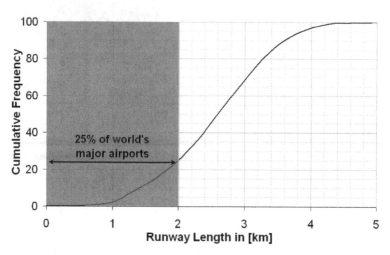

Fig. 7. Cumulative distribution of world's major runway lengths (based on data from Jenkinson et al. (1999)).

Perturbation	Exchange rate	
	Long range	Short range
1000 [kg] weight penalty	0.73%	1.26%
+1% SFC	1.28%	1.09%

Table 1. Block fuel exchange rates using the baseline long range and short range rubberised wing aircraft models.

4.2 Engine design optimality

Whereas optimisation constraints can help ensure the feasibility of an engine design, they do little to help with it's optimality. The optimality of the engine design will depend on the careful selection of the figures of merit used during the optimisation process, such as minimum block fuel, maximum time between overhaul, minimum direct operating costs, minimum noise and LTO NO_x emissions etc.

Determining the optimal aero-engine design is essentially the subject of a multi-objective optimisation, and therefore Pareto fronts need typically be constructed to visualize the region of optimal designs within the design space. A simplified example of utilizing the tool for design space exploration, with active constraints, is illustrated in Fig. 8. In principle, nacelle drag should also be added as a third dimension when plotting design space exploration results that consider varying levels of specific thrust, but this has been omitted here in order to simplify the plot. The aircraft exchange rates for the baseline design were used for plotting a constant block fuel line (ignoring nacelle drag effects and nonlinearities) and this iso-line therefore defines, in a simple manner, the boundaries of trading specific fuel consumption for weight. During a block fuel optimization, the optimizer continuously evaluates different engine designs as it searches for the optimal solution. Designs that fail to meet constraints set by the user are discarded and have been labeled as infeasible in the plot.

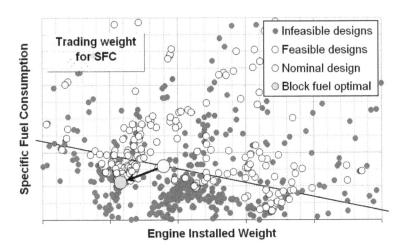

Fig. 8. Visualization example of constrained design space exploration.

4.3 Economic considerations

Safety considerations aside, civil aero engine design has been driven primarily by economic considerations even from its fairly early days. A testament to this has been the advent of the world's first commercial jet-airliner, the de Havilland Comet, powered by 4 Rolls-Royce Avon turbojet engines. Although it burned nearly four times as much fuel compared to piston-driven engines, it's business case was very strong since it permitted significantly higher flight speeds resulting in reduced flight times (i.e. a better airline product) and increased aircraft annual utilization. Furthermore, the excellent power to weight ratio of the turbojet engine meant that it could be used to power aircrafts with significantly higher passenger capacities than what was feasible before. The evolution of aircraft transport efficiency since the late 30's is summarised in Fig. 9 based on data from Avellán (2008).

The aero engine designs proposed herein have been optimized for minimum block fuel for a given aircraft mission (business case), which implies minimum global warming impact if one considers CO_2 emissions alone. The market competitiveness of these fuel optimal designs however is highly dependent on the development of jet fuel prices in the years to come until 2020. The volatility of jet fuel price over the last 10 years is illustrated in Fig. 10. A further economic consideration for European markets may also be the development of the Euro/US$ exchange rate, as well as interest and inflation rates.

For the economic calculations conducted in this study certain assumptions were made. The assumed jet fuel price was 172c$/US gallon. It is worth noting that at the time of writing the average jet fuel price was 320 [c$/US gallon] (International Air Transport Association, 2011; Platts, 2011). Interest and inflation rates were assumed to be 6% and 2%, respectively, while the US$ to Euro exchange rate was assumed to be 0.8222.

It is worth noting that an increase in inflation rates from 2% to 3% can increase the net present cost by as much as 17%, over a period of 30 years. An increase in interest rates from 6% to 7% can increase Direct Operating Costs (DOC) by 2.5% and 4.5% for short and long range applications, respectively.

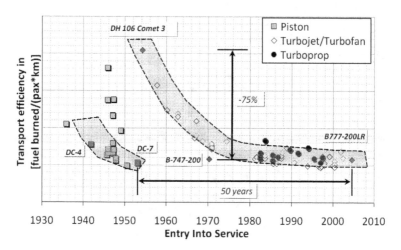

Fig. 9. Evolution of aircraft transport efficiency (based on data from Avellán (2008)).

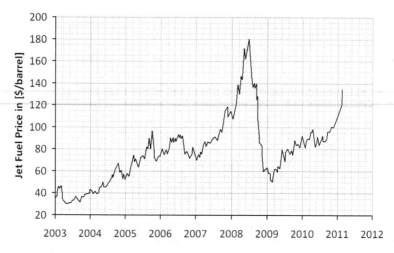

Fig. 10. Long term perspective of jet fuel price movements (based on data from International Air Transport Association (2011) and Platts (2011)).

The cost of fuel as a fraction of the total DOC was predicted to be 13% and 19% for short and long range applications, respectively. An increase in block fuel by 1% translates in an increase of 0.13% and 0.19% in DOC, respectively, and as can be observed it is directly dependent on the ratio of fuel cost over DOC. A doubling of the fuel price would change this ratio to roughly 23% and 32%, respectively, and would also result in 13% and 19% higher DOC levels, respectively.

Higher levels of DOC, as a result of a significant increase in fuel price, would most probably be absorbed by airlines through an increase in fares. This could make fuel efficient designs increasingly market competitive, as the DOC optimal designs would further approach the fuel optimal designs. It would therefore be worthwhile to redirect further research investments towards developing fuel efficient aero engine designs, as has also been the case in the late 70's and through large part of the 80's. The introduction of carbon taxes could also have a similar effect.

5. Summary of design space exploration results

A summary of three different design space exploration case studies using the tools and algorithm described is presented here. This work has looked at the potential block fuel benefits resulting from the introduction of:

- An intercooled core in a direct drive UHBR turbofan configuration.
- An intercooled recuperated core in a geared UHBR turbofan configuration.
- An open rotor propulsor in a geared pusher configuration.

The thrust requirements for the first two concepts are for an engine designed to power the long range aircraft model while the latter concept is centered around powering the short range aircraft model. More details on these studies can be found in Kyprianidis et al. (2011) and Larsson et al. (2011).

For the intercooled core assessment, a year 2020 Entry Into Service (EIS) turbofan engine with a conventional core was set up as the baseline. The intercooled core engine is an ultra high OPR design with also year 2020 EIS level of technology, and features a tubular heat-exchanger, while the fan for both engines has the same diameter and flow per unit of area. Business case block fuel benefits of approximately 3.2% are predicted for the intercooled engine, mainly due to the reduced engine weight and the core's higher thermal efficiency which results in a better SFC. **These intercooling benefits are highly dependent on achieving technology targets such as low intercooler weight and pressure losses**; the predicted lower dry weight, compared to the conventional core engine, can be attributed to various reasons. The intercooler weight penalty is largely compensated by the higher core specific output allowing a smaller core size and hence a higher BPR at a fixed thrust and fan diameter. The high OPR provides an additional sizing benefit, for components downstream of the HPC, by reducing further the corrected mass flow and hence flow areas. The intercooled core Low Pressure Turbine (LPT) was designed in this study with one less stage which reduced both engine weight and length, despite the high cycle OPR requiring a greater number of HPC stages. These observations are summarised in Table 2 with the added components weight group considering the intercooler and its installation standard; this group is not considered in the core weight group which also does not consider the core nozzle or the LPT and its casing. For the intercooled recuperated core assessment, a year 2000 EIS turbofan engine with a conventional core was set up as the baseline. The intercooled recuperated core configuration is an UHBR design with a year 2020 level of technology. Significant business case block fuel benefits of nearly 22% are predicted for the geared intercooled recuperated core engine due to its higher thermal and propulsive efficiency. The use of HPT cooling air bled from the recuperator exit (Boggia & Rud, 2005; Walsh & Fletcher, 1998) results in a 1.3% SFC improvement due to more energy being recuperated from the exhausts, at a fixed effectiveness level - and despite the considerable increase in cooling air requirements (+3.5% of core mass flow). The predicted dry weight for the intercooled recuperated configuration is higher compared to the conventional core engine. There is a weight benefit from the use of EIS

| | Conventional core | Intercooled core |
| | DDTF LR | DDIC LR |
	EIS 2020	EIS 2020
Engine dry weight	Ref.	-5.9%
LPT weight	Ref.	-27.1%
Core weight	Ref.	-32.5%
Added components weight (as % of engine dry weight)	-	7.7%
Block fuel weight	Ref.	-3.2%
Mid-cruise SFC	Ref.	-1.5%
Thermal efficiency	Ref.	+0.007
Propulsive efficiency	Ref.	+0.000

Table 2. Comparison of an intercooled engine with a conventional core turbofan engine at aircraft system level.

| | Conventional core | Intercooled recuperated core |
| | BASE LR | IRA LR |
	EIS 2000	EIS 2020
Thrust/weight	Ref.	-12%
Engine dry weight	Ref.	+16.5%
Nacelle weight	Ref.	+29.7%
Fan weight	Ref.	+36.6%
LPT weight	Ref.	-17.1%
Added components weight (as % of engine dry weight)	-	25.4%
Block fuel weight	Ref.	-21.6%
Mid-cruise SFC	Ref.	-18.3%
Thermal efficiency	Ref.	+0.024
Propulsive efficiency	Ref.	+0.120

Table 3. Comparison of an intercooled recuperated engine with a conventional core turbofan engine at aircraft system level.

2020 light-weight materials in most major engine components, as well as from the high speed LPT - due to the reduced stage count. Also, the relatively low engine OPR and the use of an intercooler increases core specific output, resulting in a smaller core. The introduction however of the gearbox, intercooler and recuperator components inevitably results in a significant weight penalty. It should be noted that a lower level of specific thrust, and hence a larger fan diameter, has been assumed for the intercooled recuperated core engine; this results in both a heavier fan and a heavier nacelle. These observations are summarised in Table 3 with the added components weight group considering the intercooler and recuperator and their installation standard, as well as the gearbox.

For the geared open rotor assessment, a year 2020 EIS geared turbofan engine with a conventional core was set up as the baseline. The geared open rotor concept design also assumes year 2020 EIS level of technology, and features two counter-rotating propellers in a pusher configuration powered by a geared low pressure turbine. Significant business case

	Geared turbofan GTF SR EIS 2020	Geared open rotor GOR SR EIS 2020
Engine installed weight	Ref.	+11%
Nacelle weight	Ref.	-88%
Fan/propeller weight	Ref.	+73%
LPT weight	Ref.	+20%
Core weight	Ref.	-31%
Block fuel weight	Ref.	-15%
Mid-cruise SFC	Ref.	-14%
Thermal efficiency	Ref.	-0.013
Propulsive efficiency	Ref.	+0.16

Table 4. Comparison of a geared open rotor engine with a geared turbofan engine at aircraft system level.

block fuel benefits of nearly 15% are predicted for the geared open rotor engine primarily due to its higher propulsive efficiency. Although, the geared turbofan engine benefits from a better thrust to weight ratio it suffers from significantly higher nacelle drag losses, compared to the open rotor design. These observations are summarised in Table 3.

A NO_x emissions assessment of the presented engine configurations has been performed and is illustrated in Fig. 11. The same combustor concept has been considered for both designs i.e., conventional Rich-burn/Quick-quench/Lean-burn (RQL) combustion technology. The results obtained are compared against ICAO Annex 16 Volume II legislative limits (ICAO, 1993), as well as the Medium Term (MT) and Long Term (LT) technology goals set by CAEP (Newton et al., 2007). Balloons have been used to indicate the uncertainty in the NO_x predictions due to the lower technology readiness level associated with the introduction of

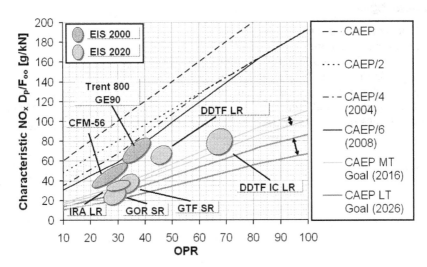

Fig. 11. NO_x emissions assessment for different future aero engine design concepts.

such combustor designs in the proposed future cycles. A sufficient margin against the ICAO CAEP/6 LTO cycle NO_x certification limit may be achieved for all the configurations that have been assessed assuming year 2020 EIS.

6. Conclusions

The research work presented started by reviewing the evolution of the aero engine industry's vision for the aero engine design of the future. Appropriate research questions were set that can influence how this vision may further involve in the years to come. Design constraints, material technology, customer requirements, noise and emissions legislation, technology risk and economic considerations and their effect on optimal concept selection were also discussed in detail.

With respect to addressing these questions, several novel engine cycles and technologies - currently under research - were identified. It was shown that there is a great potential to reduce fuel consumption for the different concepts identified, and consequently decrease the CO_2 emissions. Furthermore, this can be achieved with a sufficient margin from the ICAO NO_x certification limits, and in line with the medium term and long term goals set by CAEP. It must be noted however that aero engine design is primarily driven by economic considerations. As fuel prices increase, the impact of fuel consumption on direct operating costs also increases. The question therefore rises:

Can the potential reduction in fuel consumption and direct operating costs outweigh the technological risks involved in introducing novel concepts into the market?

The answer is left to be given by the choices the aero engine industry makes in the years to come.

7. Acknowledgements

The author is grateful to Richard Avellán (Volvo Aero) for providing the transport efficiency data used in Fig. 9. Stimulating discussions with A.M. Rolt (Rolls-Royce), J.A. Borradaile, S. Donnerhack (MTU Aero Engines), P. Pilidis, (Cranfield University), R. Singh, (Cranfield University), S.O.T. Ogaji, (Cranfield University), P. Giannakakis (Cranfield University), T. Grönstedt (Chalmers University), A. Lundbladh (Volvo Aero) and L. Larsson (Volvo Aero) on advanced concepts and aero engine design are gratefully acknowledged. Finally, the author would like to thank the reviewers of this work for their constructive suggestions to improve the overall quality and clarity of the article.

8. Nomenclature

OPR	Engine overall pressure ratio
SFC	Engine specific fuel consumption
T_4	Combustor outlet temperature

. References

Advisory Council for Aeronautical Research in Europe (2001). European Aeronautics: A Vision for 2020 – Meeting Society's Needs and Winning Global Leadership. See also URL http://www.acare4europe.org.

Avellán, R. (2008). *Towards Environmentally Friendly Aero Engines*, Licentiate thesis, Chalmers University of Technology, Göteborg, Sweden.

Bates, R. & Morris, J. (1983). A McDonnell Douglas Perspective – Commercial Aircraft for the Next Generation, *AIAA Aircraft Design, Systems and Technology Meeting Proceedings, AIAA-83-2502*, Fort Worth, Texas, USA.

Birch, N. (2000). 2020 vision: the prospects for large civil aircraft propulsion, *RAeS Aeronautical Journal* pp. 347–352.

Boggia, S. & Rud, K. (2005). Intercooled Recuperated Gas Turbine Engine Concept, *Proceedings of 41st AIAA/ASME/SAE/ASEE Joint Propulsion Conference and Exhibit, AIAA 2005-4192*, Arizona, USA.

Borradaile, J. (1988). Towards the optimum ducted UHBR engine, *Proceedings of AIAA/SAE/ASME/ASEE 24th Joint Propulsion Conference, AIAA-89-2954*, Boston, Massachusetts, USA.

Canière, H., Willcokx, A., Dick, E. & De Paepe, M. (2006). Raising cycle efficiency by intercooling in air-cooled gas turbines, *Applied Thermal Engineering* 26(16): 1780–1787.

Clean Sky Joint Technology Initiative (2011). http://www.cleansky.eu.

da Cunha Alves, M., de Franca Mendes Carneiro, H., Barbosa, J., Travieso, L., Pilidis, P. & Ramsden, K. (2001). An insight on intercooling and reheat gas turbine cycles, *Proceedings of the Institution of Mechanical Engineers, Part A: Journal of Power and Energy* 215(2): 163–171.

Downs, J. & Kenneth, K. (2009). Turbine Cooling Systems Design – Past, Present and Future, *ASME TURBO EXPO 2009 Proceedings, GT2009-59991*, Orlando, Florida.

enVIronmenTALly friendly aero engines (2009). http://www.project-vital.org.

Federal Aviation Administration (n.d.). Federal Aviation Regulation Part 121 - Operating Requirement: domestic, flag and supplemental operations, *FAR 121*, Washington, DC, USA.

GE36 Design and Systems Engineering (1987). Full-scale technology demonstration of a modern counterrotating unducted fan engine concept - engine test, *NASA-CR-180869*, GE Aircraft Engines, Cincinnati,, OH, USA.

Gray, D. & Witherspoon, J. (1976). Fuel conservative propulsion concepts for future air transports, *Technical Report SAE-760535*, Society of Automotive Engineers.

Hirschkron, R. & Neitzel, R. (1976). Alternative concepts for advanced energy conservative transport engines, *Technical Report SAE-760536*, Society of Automotive Engineers.

Horlock, J., Watson, D. & Jones, T. (2001). Limitations on Gas Turbine Performance Imposed by Large Turbine Cooling Flows, *ASME Journal of Engineering for Gas Turbines and Power* 123(3): 487–494.

ICAO (1993). International Standards and Recommended Practices - Environmental Protection, Annex 16 to the Convention on International Civil Aviation, Volume II - Aircraft Engine Emissions, *2nd edition plus ammendments*, Montreal, Canada.

International Air Transport Association (2011). http://www.iata.org.

Jackson, A. (1976). Some Future Trends in Aero Engine Design for Subsonic Transport Aircraft, *ASME Journal of Engineering for Power* 98: 281–289.

Jackson, A. (2009). *Optimisation of Aero and Industrial Gas Turbine Design for the Environment*, PhD thesis, Cranfield University, Cranfield, Bedfordshire, United Kingdom.

Jenkinson, L., Simpkin, P. & Rhodes, D. (1999). *Civil Jet Aircraft Design*, 1st edn, Arnold, London, United Kingdom.

Joint Aviation Authorities (2008). Joint Airworthiness Requirements OPS Part 1 – Commercial Air Transportation (Aeroplanes), *Ammendment 14*, Hoofddorp, The Netherlands.

Korsia, J.-J. (2009). VITAL – European R&D Programme for Greener Aero-Engines, *ISABE 2009 Proceedings, ISABE-2009-1114*, Montreal, Canada.

Korsia, J.-J. & Guy, S. (2007). VITAL – European R&D Programme for Greener Aero-Engines, *ISABE 2007 Proceedings, ISABE-2007-1118*, Beijing, China.

Kurzke, J. (2003). Achieving maximum thermal efficiency with the simple gas turbine cycle, *Proceedings of 9th CEAS European Propulsion Forum: "Virtual Engine – A Challenge for Integrated Computer Modelling"*, Rome, Italy.

Kyprianidis, K. (2010). *Multi-disciplinary Conceptual Design of Future Jet Engine Systems*, PhD thesis, Cranfield University, Cranfield, Bedfordshire, United Kingdom.

Kyprianidis, K., Colmenares Quintero, R., Pascovici, D., Ogaji, S., Pilidis, P. & Kalfas, A. (2008). EVA - A Tool for EnVironmental Assessment of Novel Propulsion Cycles, *ASME TURBO EXPO 2008 Proceedings, GT2008-50602*, Berlin, Germany.

Kyprianidis, K., Grönstedt, T., Ogaji, S., Pilidis, P. & Singh, R. (2011). Assessment of Future Aero-engine Designs with Intercooled and Intercooled Recuperated Cores, *ASME Journal of Engineering for Gas Turbines and Power* 133(1). doi:10.1115/1.4001982.

Larsson, L., Grönstedt, T. & Kyprianidis, K. (2011). Conceptual Design and Mission Analysis for a Geared Turbofan and an Open Rotor Configuration, *ASME TURBO EXPO 2011 Proceedings, GT2011-46451*, Vancouver, Canada.

Lefebvre, A. (1999). *Gas Turbine Combustion*, 2nd edn, Taylor & Francis, PA, USA.

Swiss International Air Lines, (2009). Flying Smart, *Swiss Magazine, Issue 12.2009 / 1.2010* pp. 100–105.

Lundbladh, A. & Sjunnesson, A. (2003). Heat Exchanger Weight and Efficiency Impact on Jet Engine Transport Applications, *ISABE 2003 Proceedings, ISABE-2003-1122*, Cleveland, USA.

McDonald, C., Massardo, A., Rodgers, C. & Stone, A. (2008a). Recuperated gas turbine aeroengines, part I: early development activities, *Aircraft Engineering and Aerospace Technology: An International Journal* 80(2): 139–157.

McDonald, C., Massardo, A., Rodgers, C. & Stone, A. (2008b). Recuperated gas turbine aeroengines, part II: engine design studies following early development testing, *Aircraft Engineering and Aerospace Technology: An International Journal* 80(3): 280–294.

McDonald, C., Massardo, A., Rodgers, C. & Stone, A. (2008c). Recuperated gas turbine aeroengines, part III: engine concepts for reduced emissions, lower fuel consumption, and noise abatement, *Aircraft Engineering and Aerospace Technology: An International Journal* 80(4): 408–426.

NEW Aero engine Core concepts (2011). http://www.newac.eu.

Newton, P., Holsclaw, C., Ko, M. & Ralph, M. (2007). Long Term Technology Goals for CAEP/7. presented to the Seventh Meeting of CAEP.

Papadopoulos, T. & Pilidis, P. (2000). Introduction of Intercooling in a High Bypass Jet Engine, *ASME TURBO EXPO 2000 Proceedings, 2000-GT-150*, Munich, Germany.

Peacock, N. & Sadler, J. (1992). Advanced Propulsion Systems for Large Subsonic Transports, *ASME Journal of Propulsion and Power* 8(3): 703–708.

Pellischek, G. & Kumpf, B. (1991). Compact Heat Exchanger Technology for Aero Engines, *ISABE 1991 Proceedings, ISABE-91-7019*, Nottingham, United Kingdom.

Platts (2011). http://www.platts.com.

Pope, G. (1979). Prospects for reducing the fuel consumption of civil aircraft, *RAeS Aeronautical Journal* pp. 287–295.

Rolt, A. & Baker, N. (2009). Intercooled Turbofan Engine Design and Technology Research in the EU Framework 6 NEWAC Programme, *ISABE 2009 Proceedings, ISABE-2009-1278*, Montreal, Canada.

Rolt, A. & Kyprianidis, K. (2010). Assessment of New Aero Engine Core Concepts and Technologies in the EU Framework 6 NEWAC Programme, *ICAS 2010 Congress Proceedings, Paper No. 408*, Nice, France.

Ruffles, P. (2000). The future of aircraft propulsion, *Proceedings of the IMechE, Part C: Journal of Mechanical Engineering Science* 214(1): 289–305.

Saravanamuttoo, H. (2002). The Daniel and Florence Guggenheim Memorial Lecture - Civil Propulsion; The Last 50 Years, *ICAS 2002 Congress Proceedings*, Toronto, Canada.

Saravanamuttoo, H., Rogers, G. & Cohen, H. (2001). *Gas Turbine Theory*, 5th edn, Pearson Education Limited, United Kingdom.

Schimming, P. (2003). Counter Rotating Fans – An Aircraft Propulsion for the Future, *Journal of Thermal Science* 12(2): 97–103.

Schoenenborn, H., Ebert, E., Simon, B. & Storm, P. (2006). Thermomechanical Design of a Heat Exchanger for a Recuperated Aeroengine, *ASME Journal of Engineering for Gas Turbines and Power* 128(4): 736–744.

Sieber, J. (1991). Aerodynamic Design and Experimental Verification of an Advanced Counter-Rotating Fan for UHB Engines, *Third European Propulsion Forum*, Paris, France.

Swihart, J. (1970). The Promise of the Supersonics, *AIAA 7th Annual Meeting and Technical Display Proceedings, AIAA 70-1217*, Houston, Texas, USA.

valiDation of Radical Engine Architecture systeMs (2011). http://www.dream-project.eu.

Walker, A., Carrotte, J. & Rolt, A. (2009). Duct Aerodynamics for Intercooled Aero Gas Turbines: Constraints, Concepts and Design Methododology, *ASME TURBO EXPO 2009 Proceedings, GT2009-59612*, Orlando, Florida.

Walsh, P. & Fletcher, P. (1998). *Gas Turbine Performance*, 1st edn, Blackwell Science, United Kingdom.

Watts, R. (1978). European air transport up to the year 2000, *RAeS Aeronautical Journal* pp. 300–312.

Wilcock, R., Young, J. & Horlock, J. (2005). The Effect of Turbine Blade Cooling on the Cycle Efficiency of Gas Turbine Power Cycles, *ASME Journal of Engineering for Gas Turbines and Power* 127(1): 109–120.

Wilde, G. (1978). Future large civil turbofans and powerplants, *RAeS Aeronautical Journal* 82: 281–299.

Wilfert, G., Sieber, J., Rolt, A., Baker, N., Touyeras, A. & Colantuoni, S. (2007). New Environmental Friendly Aero Engine Core Concepts, *ISABE 2007 Proceedings*, *ISABE-2007-1120*, Beijing, China.

Xu, L. & Grönstedt, T. (2010). Design and Analysis of an Intercooled Turbofan Engine, *ASME Journal of Engineering for Gas Turbines and Power* 132(11). doi:10.1115/1.4000857.

Xu, L., Gustafsson, B. & Grönstedt, T. (2007). Mission Optimization of an Intercooled Turbofan Engine, *ISABE 2007 Proceedings*, *ISABE-2007-1157*, Beijing, China.

Young, P. (1979). The future shape of medium and long-range civil engines, *RAeS Aeronautical Journal* pp. 53–61.

Zimbrick, R. & Colehour, J. (1988). An investigation of very high bypass ratio engines for subsonic transports, *Proceedings of AIAA/SAE/ASME/ASEE 24th Joint Propulsion Conference*, *AIAA-88-2953*, Boston, Massachusetts, USA.

2

Possible Efficiency Increasing of Ship Propulsion and Marine Power Plant with the System Combined of Marine Diesel Engine, Gas Turbine and Steam Turbine

Marek Dzida
Gdansk University of Technology
Poland

1. Introduction

For years there has been, and still is, a tendency in the national economy to increase the efficiency of both the marine and inland propulsion systems. It is driven by economic motivations (rapid increase of fuel prices) and ecological aspects (the lower the fuel consumption, the lower the emission of noxious substances to the atmosphere). New design solutions are searched to increase the efficiency of the propulsion system via linking Diesel engines with other heat engines, such as gas and steam turbines. The combined systems implemented in marine propulsion systems in recent years are based mainly on gas and steam turbines (MAN, 2010). These systems can reach the efficiency exceeding 60% in inland applications. The first marine system of this type was applied on the passenger liner "Millenium". However, this is the only high-efficiency marine application of the combined propulsion system so far. Its disadvantage is that the system needs more expensive fuel, the marine Diesel oil, while the overwhelming majority of the merchant ships are driven by low-speed engines fed with relatively cheap heavy fuel oil. It seems that the above tendency will continue in the world's merchant navy for the next couple of years.

The compression-ignition engine (Diesel engine) is still most frequently used as the main engine in marine applications. It burns the cheapest heavy fuel oil and reveals the highest efficiency of all heat engines. The exhaust gas leaving the Diesel engine contains huge energy which can be utilised in another device (engine), thus increasing the efficiency of the entire system and reducing the emission of noxious substances to the atmosphere.

A possible solution here can be a system combined of a piston internal combustion engine and the gas and steam turbine circuit that utilises the heat contained in the exhaust gas from the Diesel engine. The leading engine in this system is the piston internal combustion engine. It seems that now, when fast container ships with transporting capacity of 8- 12 thousand TU are entering into service, the propulsion engines require very large power, exceeding 50-80 MW. On the other hand, increasing prices of fuel and restrictive ecological limits concerning the emission of NO_x and CO_2 to the atmosphere provoke the search for new solutions which will increase the efficiency of the propulsion and reduce the emission of gases to the atmosphere.

The ship main engines will be large low-speed piston engines that burn heavy fuel oil. At present, the efficiency of these engines nears 45 – 50%. For such a large power output ranges, the exhaust gas leaving the engine contains huge amount of heat available for further utilisation.

The proposed combined system consisting of a piston internal combustion engine, a gas turbine and a steam turbine can also be used for engines of lower power, ranging between 400 ÷900 kW. For those power ranges a use of low-boiling media of organic-based refrigerant type instead of water (steam) in the steam cycle seems to be a reasonable solution. Piston internal combustion engines of this power range are used on coasting vessels, or in the inland water transport, for instance for driving cargo barges. On the other hand in inland applications the power blocks fired with solid, liquid, or gas fuels are in almost 100% the systems with steam or gas turbines.

In the Central Europe, Poland for instance, the basic fuel in power engineering is coal. Conventional electric power plants have the efficiency of an order of 38-42%, and emit large volumes of CO_2 , NO_x and/or SO_x. In order to decrease the amount of noxious substances emitted to the atmosphere and reduce the cost of production of the electric energy, combined systems are in use - consisting of gas turbines with a steam turbine circuit.

On the other hand, the combined turbine power plants can be complemented by electric power plants with a Diesel engine as the main propulsion. The exhaust gas leaving the engine contains about 30-40% of the heat delivered to the engine in the fuel. Using the heat from the exhaust gas in the gas and steam turbine circuit will increase the efficiency of the entire combined system. For large powers of piston internal combustion engines, the additional gas and steam turbine circuit is a source of measurable economic savings in electric energy production. Moreover, in large-power piston internal combustion engines we can additionally use the low-temperature waste heat, for instance for heating the communal water (Dzida, 2009). In the seaside areas with no large electric power plants, a mobile power plant situated on a platform close to the coast reveals additional advantages:

- increasing production of electric energy in the so-called distributed system,
- diversification of primary energy sources which decreases the consumption of coal in favour of liquid fuels,
- possible combustion of residual heavy fuels from nearby oil refineries,
- reducing large-distance transport of solid fuels, the absence of slag and cinders,
- reducing the emission of CO_2 and NO_x due to the increased system efficiency,
- shorter time of plant erection compared to that of a conventional power plant, and possibility of opening it in stages: first with the Diesel engine alone, and then complementing it, during plant operation, with a combined steam/gas turbine system,
- no problems with the water cooling the condenser, small effect on the environment in water balance aspects,
- mobility of a combined power plant erected on the marine platform.

2. Concept of a combined system

Combined propulsion systems are used in marine engineering mostly in fast special-purpose ships and in the Navy, as the systems being a combination of a Diesel engine and gas turbines (CODAG, CODOG) or solely gas turbines (COGOG, COGAG). The propulsion system of the passenger liner "Millenium" uses a COGES-type system which improved the efficiency and operating abilities of the ship. The system consists of a gas turbine and a steam turbine which drive an electric current generator, while the propeller screws are driven by electric motors. In this system the steam turbine circuit is supplied with the steam generated in the waste heat boiler supplied with the exhaust gas from the gas turbines.

Possible Efficiency Increasing of Ship Propulsion and Marine Power Plant with the System Combined of Marine Diesel Engine, Gas Turbine and Steam Turbine

27

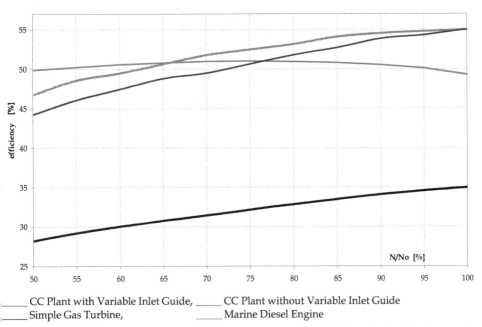

_____ CC Plant with Variable Inlet Guide, _____ CC Plant without Variable Inlet Guide
_____ Simple Gas Turbine, _____ Marine Diesel Engine

Fig. 1. Part - load Efficiency of a Combined - Cycle Plant (GT&ST), Simple Gas Turbine and Marine Diesel Engine

Combined systems used in inland power blocks base on a gas turbine as the main unit and a steam turbine that utilises the steam produced in a waste heat boiler using the heat recovered from the gas turbine exhaust gas. All this provides opportunities for reaching high efficiency of the combined block. The exhaust gas leaving a marine low-speed Diesel engine contains smaller amount of heat, of an order of 30-40% of the energy delivered to the engine.

Figure 1 shows sample efficiency curves of the combined gas turbine/steam turbine systems as functions of power plant load, compared to the gas turbine operating in a simple open circuit and the marine low-speed Diesel engine.

The efficiency curves in Fig. 1 show that the combined cycle gas turbine/steam turbine system has the highest efficiency for maximal loads (maximal efficiency levels for these circuits reach as much as 60%). Gas turbines operating in the simple open circuit have the lowest efficiency (average values of 33÷35%, and maximal values reaching 40%). Low-speed Diesel engines have the efficiency of an order of 47÷50%. It is also noticeable that the Diesel engine curve is relatively flat. This is of special importance in case of marine propulsion systems which operate at heavily changing loads. For the combined cycle gas turbine/steam turbine systems and the gas turbines operating in the simple open circuit the relative efficiency decrease $\Delta\eta/\eta_0$ is equal to 15÷20% when the load decreases from 100% to 50%. For the low-speed Diesel engine these numbers are equal to 1÷2%. This property of the Diesel engine, along with the ability to utilise additional heat contained in its exhaust gas, makes the engine the most applicable in marine propulsion systems operating in heavily changing load conditions. The amount of heat contained in the exhaust gas from the gas turbine is approximately equal to 60÷65%, i.e. more than in piston engines, which results from lower exit temperature and less intensive flow of the exit gas leaving the Diesel engine, Figs. 2 and 3.

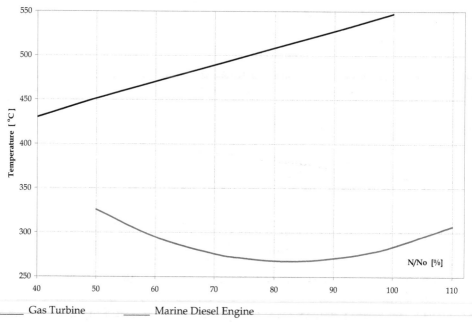

_____ Gas Turbine   _____ Marine Diesel Engine

Fig. 2. Temperatures of the exhaust gas from the Diesel engine and the gas turbine as function of power plant load

The exit temperatures of the exhaust gas from the gas turbines range between 450÷600°C, on average, while those from the low-speed Diesel engines are of an order of 220÷300°C. In the gas turbines, decreasing the load remarkably decreases the temperature of the exhaust gas, while in the Diesel engine these changes are much smaller, and the temperature initially decreases and then starts to increase for low loads.

This property of the steam turbine circuit in the combined system with the Diesel engine for partial loads makes it possible to keep the live steam temperature at a constant level within a wide range of load. The related exhaust gas mass flow rate m_g/N [kg/kWh] changes only by about 5% in the Diesel engine when the load changes from 100% to 50%, while in the gas turbine this parameter changes by about 55% for the same load change, Fig. 3.

The combined propulsion system with the low-speed piston internal combustion engine used as the main engine and making use of the heat from the engine exhaust gas is shown in Fig. 4, (Dzida, 2009; Dzida & Mucharski, 2009; Dzida et al., 2009).

The exhaust gas flows leaving individual main engine cylinders are collected in the exhaust manifold and passed to the constant-pressure turbocharger. Due to high turbocharger efficiency ranges (MAN, 2010; Schrott, 1995), the scavenge air can be compressed using the energy contained only in part of the exhaust gas flow. The remaining part of the exhaust gas flow can be expanded in an additional gas turbine, the so-called power turbine, which additionally drives, via a gear, the propeller screw or the electric current generator.

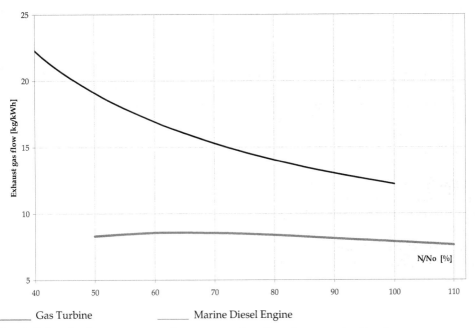

_____ Gas Turbine   _____ Marine Diesel Engine

Fig. 3. Related exhaust gas mass flow rate as a function of power plant load

The exhaust gas from the turbocharger and the power turbine flows to the waste heat boiler installed in the main engine exhaust gas path, before the silencer. The waste heat boiler produces the steam used both for driving the steam turbine that passes its energy to the propeller screw, and for covering all-ship needs.

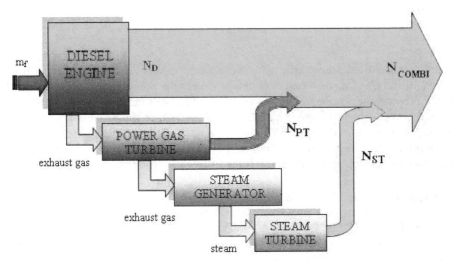

Fig. 4. Concept of the combined propulsion system

In the marine low-speed Diesel engines, another portion of energy that can be used along with the exhaust gas energy is a huge amount of so-called waste heat of relatively low temperature. In the low-speed engines the waste heat comprises the following components (with their proportions to the heat delivered to the engine in fuel):

- heat in the scavenge air cooler (17-20%), of an approximate temperature of about 200°C,
- heat in the lubricating oil cooler (3-5%), of an approximate temperature of about 50⁰.
- heat in the jacket water cooler (5-6%), of the temperature of an order of 100°C.

This shows that the amount of the waste heat that remains for our disposal is equal to about 25-30% of the heat delivered in fuel. Part of this heat can be used in the combined circuit with the Diesel engine.

2.1 Energy evaluation of the combined propulsion system

The adopted concept of the combined ship propulsion system requires energy evaluation, Fig. 4. Formulas defining the system efficiency are derived on the basis of the adopted scheme.

The power of the combined propulsion system is determined by summing up individual powers of system components (the main engine, the power gas turbine, and the steam turbine):

$$N_{combi} = N_D + N_{PT} + N_{ST} \tag{1}$$

hence the efficiency of the combined system is:

$$\eta_{combi} = \frac{N_{combi}}{m_{fD} \cdot Wu} = \eta_D \cdot \left(1 + \frac{N_{PT}}{N_D} + \frac{N_{ST}}{N_D} \right) \tag{2}$$

and the specific fuel consumption is:

$$b_{ecombi} = b_{eD} \cdot \frac{1}{(1 + \frac{N_{PT}}{N_D} + \frac{N_{ST}}{N_D})} \; [g / kWh] \tag{3}$$

where η_D, b_{eD}- is the efficiency and specific fuel consumption of the main engine.

Relations (2) and (3) show that each additional power in the propulsion system increases the system efficiency and, consequently, decreases the fuel consumption. And the higher the additional power achieved from the utilisation of the heat in the exhaust gas leaving the main engine, the lower the specific fuel consumption. Therefore the maximal available power levels are to be achieved from both the power gas turbine and the steam turbine. The power of the steam turbine mainly depends on the live steam and condenser parameters.

2.2 Variants of the combined ship propulsion systems or marine power plants

For large powers of low-speed engines, the exhaust gas leaving the engine contains huge amount of heat available for further utilisation. Marine Diesel engines are always supercharged. Portions of the exhaust gas leaving individual cylinders are collected in the exhaust gas collector, where the exhaust gas pressure $p_{exh\_D} > p_{bar}$ is equalised. In standard solutions the constant-pressure turbocharger is supplied with the exhaust gas from the exhaust manifold to generate the flow of the scavenge air for supercharging the internal combustion engine.

Present-day designs of turbochargers used in piston engines do not need large amounts of exhaust gas, therefore it seems reasonable to use a power gas turbine complementing the operation of the steam turbine in those cases. Here, two variants of power gas turbine supply with the exhaust gas are possible.

2.2.1 Parallel power gas turbine supply (variant A)

In this case part of the exhaust gas from the piston engine exhaust manifold supplies the Diesel engine turbocharger. The remaining part of the exhaust gas from the manifold is directed to the gas turbine, bearing the name of the power turbine (PT). The power turbine drives, via the reduction gear, the propeller screw or the electric current generator, thus additionally increasing the power of the entire system. Figure 5 shows a concept of this propulsion system, referred to as parallel power turbine supply. After the expansion in the turbocharger and the power turbine, the exhaust gas flowing from these two turbines is directed to the waste heat boiler in the steam circuit.

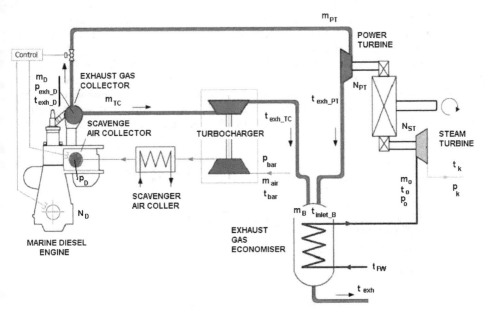

Fig. 5. Combined system with the Diesel main engine, the power turbine supplied in parallel, and the steam turbine (variant A)

In the proposed solution, at low load ranges the amount of the exhaust gas from the main engine is not sufficient to additionally supply the power turbine. In such case a control valve closes the exhaust gas flow to the power turbine, Figure 5. The operation of this valve is controlled by the control system using two signals: the scavenge air pressure signal, and the signal of the propeller shaft angular speed or torque. The waste heat boiler produces the steam which is then used both in the steam turbine and, in case of marine application, to cover the all-ship needs. This system allows for independent operation of the Diesel engine, with the steam turbine or the power turbine switched off. The control system makes it possible to switch off the power turbine thus increasing the power of the turbocharger at partial load, and, on the other hand, direct part of the Diesel engine exhaust gas to supply the power turbine at large load.

Power turbine calculations are based on the Diesel engine parameters, i.e. the temperature of the exhaust gas in the exhaust gas collector, which in turn depends on the engine load and air parameters at the engine inlet. Marine engine producers most often deliver the data on two reference points for the atmospheric air (the ambient reference conditions):

	ISO Conditions	Tropical Conditions
Ambient air temperature [^{0}C]	25	45
Barometric pressure [bar]	1	1

2.2.2 Series power gas turbine supply (variant B)

In this variant the exhaust gas from the exhaust manifold supplies first the piston engine turbocharger and then the power turbine, Fig.6.

After leaving the exhaust manifold, the exhaust gas expands in the turbocharger to the higher pressure than the atmospheric pressure, which leaves part of the exhaust gas enthalpy drop for utilisation in the power turbine. The exhaust gas leaving the power turbine passes its heat to the steam in the waste heat boiler, thus producing additional power in the steam turbine circuit.

Also in this combined system, the installed control valve makes it possible to switch off the power turbine at partial piston engine loads, thus increasing the power of the turbocharger by expanding the exhaust gas to lower pressure, Fig. 6. Unlike the parallel supply variant, here the entire mass of the exhaust gas from the piston engine manifold flows through the turbocharger. The exhaust gas pressure at the turbocharger outlet is higher than in variant A.

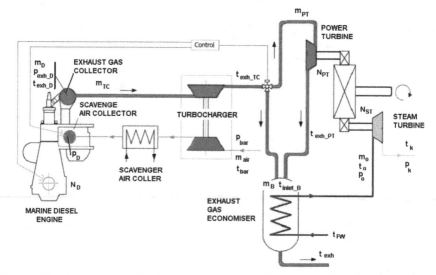

Fig. 6. Combined system with the Diesel main engine, the power turbine supplied in series, and the steam turbine (variant B)

3. Power turbine in the combined system

Calculating the power turbine in the combined system depends on the selected variant of power turbine supply. Usually, piston engine producers do not deliver the exhaust gas temperature in the exhaust manifold (which is equal to the exhaust gas temperature at turbocharger turbine inlet). Instead, they give the exhaust gas temperature at turbocharger turbine outlet (t_{exh_D}). The temperatures of the exhaust gas in the Diesel engine exhaust gas collector are calculated from the turbine power balance, according to the following formula:

$$t_{exh\_D} = \frac{t_{exh\_TC} + 273,15}{\left(1 - \eta_T \cdot \left(1 - \dfrac{1}{\pi_T^{\frac{\kappa_g - 1}{\kappa_g}}}\right)\right)} - 273,15 \ [°C] \tag{4}$$

This formula needs the data on turbocharger turbine efficiency changes for partial loads. These data can be obtained from the producer of the turbocharger (as they are rarely made public), Fig. 7, or calculated based on the relation used in steam turbine stage calculations:

$$\overline{\eta_T} \equiv \frac{\eta_T}{\eta_{To}} = 2 \cdot \overline{v} - \overline{v}^2 \tag{5}$$

where v - related turbine speed indicator, η_{To}- maximal turbine efficiency and the corresponding speed indicator.

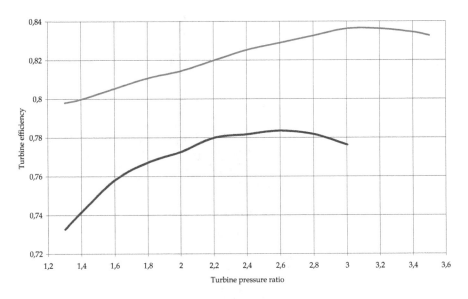

_____ turbine of S – wheel type   _____ turbine of R – wheel type

Fig. 7. Turbocharger turbine efficiency as a function of scavenge air pressure, acc. to (Schrott, 1995)

The turbine speed indicator is defined as:

$$v = \frac{u}{c_s} = \sqrt{\frac{u^2}{2 \cdot H_T}} \tag{6}$$

where u- circumferential velocity on the turbine stage pitch diameter, H_T- enthalpy drop in the turbine.

The calculations make use of static characteristics of the turbocharger compressor, with the marked line of cooperation with the Diesel engine, Fig.8.

Figure 9 shows the turbocharger efficiency curves calculated from the relation:

$$\eta_{TC} = \eta_T \cdot \eta_C \cdot \eta_m \tag{7}$$

where η_T - the turbocharger turbine efficiency is calculated from relation (5), while the compressor efficiency η_C is calculated from the line of Diesel engine/compressor cooperation, η_m – mechanical efficiency of the turbocharger, Fig. 8. In the same figure a comparison is made between the calculated turbocharger turbine efficiency with the producer's data as a function of the Diesel engine scavenge pressure. The differences between these curves do not exceed 1,5%.

For the presently available turbocharger efficiency ranges, the amount of the exhaust gas needed for driving the turbocharger turbine is smaller than the entire mass flow rate of the exhaust gas leaving the Diesel engine. Fig. 10 shows sample curves of exhaust gas

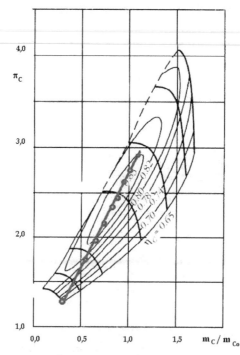

Fig. 8. Diesel engine cooperation line against turbocharger compressor characteristics

Possible Efficiency Increasing of Ship Propulsion and Marine Power Plant with the System Combined of Marine Diesel
Engine, Gas Turbine and Steam Turbine

35

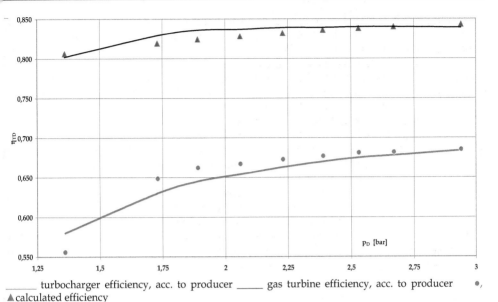

_____ turbocharger efficiency, acc. to producer _____ gas turbine efficiency, acc. to producer ●,
▲ calculated efficiency

Fig. 9. Efficiency characteristics of the turbocharger and the turbocharger gas turbine as a function of scavenge air pressure

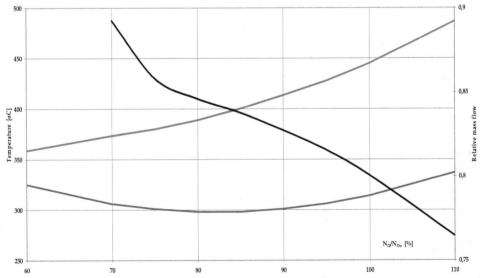

_____ temperature in the Diesel engine exhaust gas collector-calculated curves _____ exhaust gas temperature at turbocharger outlet – producer's data _____ Diesel engine exhaust gas mass flow rate related to the scavenge air mass flow rate

Fig. 10. Sample temperature characteristics of the turbocharger during gas expansion in the turbine to the atmospheric pressure and the related exhaust gas mass flow rates as functions of Diesel engine load

temperature changes in the engine manifold (calculated using the relation (4)) and the exhaust gas temperature at the turbocharger outlet (according to the data delivered by the producer) as functions of engine load, when the standard internal combustion engine exhaust gas is expanded to the barometric pressure. The figure also shows the Diesel engine exhaust gas flow rate related to the scavenge air flow rate, as a function of the engine load. This high efficiency of the turbocharger provides opportunities for installing a power gas turbine connected in parallel with the turbocharger (variant A).

The turbocharger power balance indicates that in the power gas turbine we can utilise between 10 and 24% of the flow rate of the exhaust gas leaving the exhaust manifold of the piston engine. The power gas turbine can be switched on when the main engine power output exceeds 60%. For lower power outputs the entire exhaust gas flow leaving the Diesel engine is to be used for driving the turbocharger.

In variant B of the combined system with the power turbine, the turbocharger is connected in series with the power gas turbine. Here, the entire amount of the exhaust gas flows through the turbocharger turbine. Due to the excess of the power needed for driving the turbocharger, the final expansion pressure at turbocharger turbine output can be higher than the exhaust gas pressure at waste heat boiler inlet. In this case the expansion ratio in the turbocharger turbine is given by the relation:

$$\pi_T = \left[\cfrac{1}{1 - \cfrac{1}{\eta_{TC}} \cdot \cfrac{m_a}{m_D} \cdot \cfrac{c_a}{c_g} \cdot \cfrac{t_a}{t_{exh\_D}} \cdot \left(\pi_C^{\frac{\varkappa_a - 1}{\varkappa_a}} \right)} \right]^{\frac{\varkappa_g}{\varkappa_g - 1}} \tag{8}$$

where: π_C- compression ratio of the turbocharger compressor.

The exhaust gas temperature at turbocharger outlet is calculated from the formula:

$$t_{exh\_TC} = \left(t_{exh\_D} + 273,15 \right) \cdot \left[1 - \eta_T \left(1 - \cfrac{1}{\pi_T^{\frac{\varkappa_g - 1}{\varkappa_g}}} \right) \right] - 273,15 \; [^\circ C] \tag{9}$$

Figure 11 shows sample curves of temperature, compression and expansion rate changes in the turbocharger for variant B: series power turbine supply.

This case provides opportunities for utilising the enthalpy drop of the expanding exhaust gas in the power turbine. The operation of the power turbine is possible when the Diesel engine power exceeds 60%.

3.1 Power turbine in parallel supply system (variant A)

The power turbine (Fig.5) is supplied with the exhaust gas from the exhaust manifold. The exhaust gas mass flow rate m_{PT} and temperature t_{exh_D} are identical as those at turbocharger outlet: the mass flow rate of the exhaust gas flowing through the power turbine results from the difference between the mass flow rate of the Diesel engine exhaust gas and of that expanding in the turbocharger:

$$m_{TD} = m_a \cdot (1 - \overline{m}) + m_{fD} \tag{10}$$

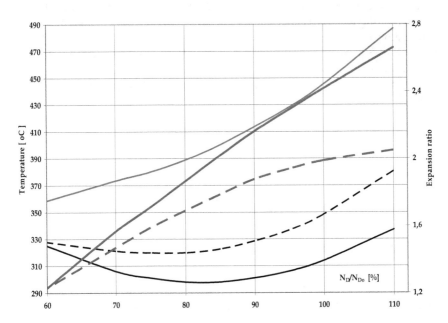

_____expansion ratio in the turbocharger turbine (standard arrangement - without power turbine) _ _
_ expansion ratio in the turbocharger turbine with power turbine _____exhaust gas temperature in the
Diesel engine exhaust gas collector ____exhaust gas temperature at turbocharger outlet without power
turbine _ _ _ exhaust gas temperature at turbocharger outlet with power turbine

Fig. 11. Changes of temperature and expansion ratio of the turbocharger in the combined
system with series power turbine supply (variant B)

The mass flow rate of the exhaust gas needed by the turbocharger is calculated from the
turbocharger power balance using the following formula:

$$\overline{m} \equiv \frac{m_{TC}}{m_a} = \frac{\pi_T^{\frac{\kappa_g-1}{\kappa_g}}}{\pi_C^{\frac{\kappa_a-1}{\kappa_a}} - 1} \cdot \frac{T_{exh\_D}}{T_a} \cdot \frac{c_g}{c_p} \cdot \eta_{TC} \tag{10.1}$$

The exhaust gas expanding in the power turbine has the inlet and outlet pressures identical
to those of the exhaust gas flowing through the turbocharger. The power of the power
turbine is given by the relation:

$$N_{PT} = \eta_m \cdot \eta_{PT} \cdot m_{PT} \cdot H_{PT} \tag{11}$$

where η_m- mechanical efficiency of the power turbine, H_{PT} – iso-entropic enthalpy drop in
the power turbine.
The power turbine efficiency η_{PT} is assumed in the same way as for the turbocharger
turbine, Fig. 9, or using the relation (5). In the shipbuilding, the gas turbines used in
combined Diesel engine systems with power turbines are those adopted from turbochargers.

The power turbine system calculations show that the exhaust gas temperature at the power turbine outlet is slightly higher than that at the turbocharger outlet, Fig.12. The increase of the main engine load results in the increase of both the exhaust gas temperature in the exhaust gas collector and the mass flow rate of the exhaust gas flowing through the power turbine. The increase in power of the combined system with additional power turbine ranges from about 2% for Diesel engine loads of an order of 70% up to over 8% for maximal loads, Fig.12.

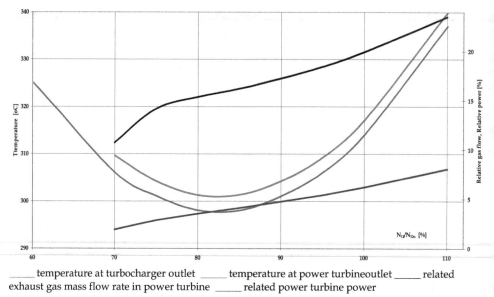

_____ temperature at turbocharger outlet _____ temperature at power turbineoutlet _____ related exhaust gas mass flow rate in power turbine _____ related power turbine power

Fig. 12. Parameters of parallel supplied power turbine as functions of the main engine load – variant A (calculations for tropical conditions)

When the Diesel engine power is lower than 60-70% of the nominal value the entire exhaust gas flow from the exhaust manifold is directed to the turbocharger drive. In this case the control system closes the valve controlling the exhaust gas flow to the power turbine, Fig. 5.

3.2 Power turbine in series supply system (variant B)
In this variant the power turbine is supplied with the full amount of the exhaust gas leaving the Diesel engine exhaust manifold. The power turbine is installed after the turbocharger. The exhaust gas pressure at the power turbine inlet depends on the pressure of the exhaust gas leaving the turbocharger turbine, Fig.11.
In this case the power of the power turbine is calculated as:

$$N_{PT} = \eta_{PT} \cdot m_D \cdot c_g \cdot t_{inl\_PT} \cdot \left(1 - \frac{1}{\pi_{PT}^{\frac{N_g-1}{N_g}}} \right) \qquad (12)$$

where t_{inl_PT} - exhaust gas temperature at the power turbine inlet, π_{PT}– expansion ratio in the power turbine , η_{PT}– power turbine efficiency. The power turbine efficiency is assumed in the same way as in variant A.
In formula (12) the exhaust gas temperature at the power turbine inlet is assumed equal to that of the exhaust gas leaving the turbocharger, Fig. 13.
The exhaust gas temperature at the power turbine output is calculated from the formula:

$$t_{exh\_PT} = \left(t_{inl\_PT} + 273,15\right)\cdot\left[1-\eta_{PT}\left(1-\frac{1}{\pi_{PT}^{\frac{\aleph g-1}{\aleph g}}}\right)\right] - 273,15[°C] \qquad (13)$$

Figure 13 also shows the expansion ratio, the power of the power turbine, and the exhaust gas temperatures at the turbocharger and the power turbine outlets for partial engine loads. The power turbine in this variant increases the power of the combined system by 3% to 9% with respect to that of a standard engine. The turbine power increases with increasing Diesel engine load.

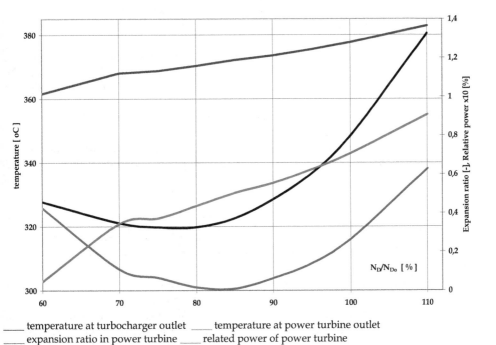

_____ temperature at turbocharger outlet _____ temperature at power turbine outlet
_____ expansion ratio in power turbine _____ related power of power turbine

Fig. 13. Parameters of series supplied power turbine as functions of the main engine load - variant B (calculations for tropical conditions)

3.3 Comparing the two power turbine supply variants
The analysis of the two examined variants shows that the power of the combined system increases depending on the Diesel engine load. For both variants the power turbine can be

used after exceeding about 65% of the Diesel engine power. The exhaust gas leaving the power turbine is directed to the waste heat boiler, where together with steam turbine it can additionally increase the overall power of the combined system.

In both cases the temperatures of the exhaust gas leaving the power turbine are comparable. The exhaust gas pressure at power turbine outlet depends on the losses generated when the gas flows through the waste heat boiler and outlet silencers. Following practical experience, the exhaust gas back pressure is assumed higher than the barometric pressure by 300 mmWC, i.e. about 3%. Taking into account powers of the power turbines for the above variants, Fig. 14, it shows that for the same Diesel engine parameters the series supply of the power turbine results in higher turbine power. For lower loads, the power of the series supplied power turbine increases, compared to the parallel supply variant.

4. Steam turbine circuit

The combined system makes use of the waste heat from the Diesel engine. In modern Diesel engines the temperatures of the waste heat are at the advantageous levels for the steam turbine circuit. This circuit makes use of water that can be utilised in a low-temperature process. Adding the steam circuit to the combined Diesel engine/power gas turbine system provides good opportunities for increasing the power of the combined system, and consequently, also the system efficiency, see formula (2).

In the examined combined system the exhaust gas leaving the turbocharger and the power turbine (variant A, Fig. 5) or only the power turbine (variant B, Fig. 6) flows to the waste heat boiler where it is used for producing superheated steam for driving the steam turbine.

The mass flow rate of the exhaust gas reaching the waste heat boiler is equal to that leaving the Diesel engine exhaust gas collector. The exhaust gas temperature at waste heat boiler inlet depends on the adopted solution of power turbine supply. For variant A with parallel supply it is calculated from the balance of mixing of the gases leaving the turbocharger and the power turbine:

$$t_{inl\_B} = \frac{m_{TC} \cdot i_{exh\_TC} + m_{PT} \cdot i_{exh\_PT}}{m_D \cdot c_g} - 273,15 \ [^{\circ}C] \tag{14}$$

while for the series power turbine supply (variant B) it is assumed equal to that at the power turbine outlet, formula (13).

In combined steam turbine systems for small power ranges and low live steam temperatures the single pressure systems are used, Fig. 15, (Kehlhofer, 1991).

Such system consists of a single-pressure waste heat boiler, a condensing steam turbine, a water-cooled condenser, and a single stage feed water preheater in the deaerator.

The main disadvantage of the systems of this type is poor utilisation of the heat contained in the exhaust gas (the waste heat energy). The steam superheater is relatively large, as the entire mass of the steam produced by the boiler flows through it. However, costs of this steam system are the lowest, as poor utilisation of the exhaust gas energy results in high temperature of the exhaust gas leaving the boiler. The deaerator is supplied with the steam extracted from the steam turbine. The application of the single pressure system does not secure optimal utilisation of the exhaust gas energy.

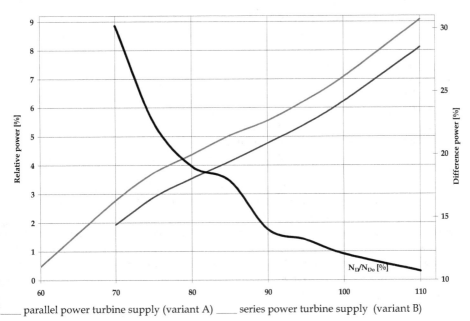

_____ parallel power turbine supply (variant A) _____ series power turbine supply (variant B)

Fig. 14. Powers of the power turbine as functions of main engine load

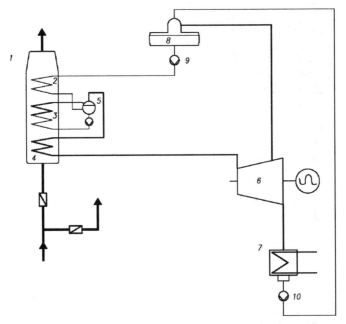

1-Waste Heat Boiler 2-Superheater 3- Evaporator 4-Ekonomizer 5-Boiler drum 6-Steam turbine 7-Condenser 8-Deaerator 9-Feed water pump 10-Condensate pump

Fig. 15. Flow Diagram of the Single Pressure System

Those steam turbine systems frequently make use of an additional low-pressure evaporator, Fig. 16, which leads not only to more intensive utilisation of the waste heat contained in the exhaust gas, but also to better thermodynamic use of the low-pressure steam.

In this solution the high pressure superheater is relatively small, compared to the single pressure boiler. The deaerator is heated with the saturated steam from the low-pressure evaporator. The power of the main high-pressure feeding pump is also smaller. The excess steam from the low-pressure evaporator can be used for supplying the low-pressure part of the steam turbine, thus increasing its power, or, alternatively, for covering all-ship needs.

Figure 16 shows possible use of the temperature waste heat from the scavenge air cooler, the lubricating oil cooler, and from the jacket water cooler in the low-pressure water pre-heater.

The additional low-pressure exchanger in the steam circuit, Fig. 16, makes it possible to increase the temperature of the water in the deaerator. Higher water temperature is required due to the presence of sulphur in the fuel (water dew-point in the exhaust gas) – it is favourable for systems fed with a high sulphur content fuel. If the temperature of the feedwater is low when the system is fed with fuel without sulphur, the heat exchanger 14 in Fig. 16 is not necessary and the waste heat from the coolers can be used in the deaerator. For a low feedwater temperature the deaerator works at the pressure below atmospheric (under the vacuum).

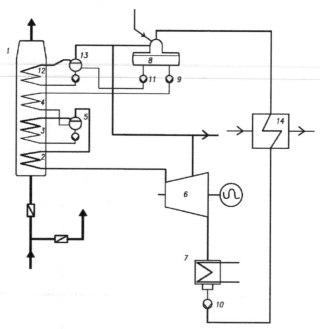

1-Waste Heat Boiler 2-High pressure superheater 3- High pressure evaporator 4- High pressure economizer 5- High pressure boiler drum 6 -Steam turbine 7-Condenser 8- Deaerator 9-High pressure feed water pump 10-Condensate pump 11-Low pressure feed pump 12-Low pressure evaporator 13-Low pressure boiler drum 14-Low pressure pre-heater

Fig. 16. Flow Diagram for a Two – Pressure System

4.1 Limits for steam circuit parameters

The limits for the values of the steam circuit parameters result from strength and technical requirements concerning the durability of particular system components, but also from design and economic restrictions. The difference between the exhaust gas temperature and the live steam temperature, Δt, for waste heat boilers used in shipbuilding is assumed as Δt = 10-15°C, according to (MAN, 1985; Kehlhofer, 1991). The "pitch point" value recommended by MAN B&W (MAN, 1985) for marine boilers is δt = 8-12°C. The limiting dryness factor x of the steam downstream of the steam turbine is assumed as x_{limit}=0,86-0,88. For marine condensers cooled with sea water, MAN recommends the condenser pressure p_K=0,065 bar. This pressure depends on the B&W (MAN, 1985) temperature of the cooling medium in the condenser. Figure 17 shows the dependence of the condenser pressure on the cooling medium temperature. The temperature of the boiler feed water is of high importance for the life time of the feed water heater in the boiler. The value of this temperature is connected with a so-called exhaust gas dew-point temperature. Below this temperature the water condensates on heater tubes and reacts with the sulphur trioxide SO_3 producing the sulphuric acid, which is the source of low-temperature corrosion. That is why boiler producers give minimal feed water temperatures below which boiler operation is highly not recommended. The dew-point temperature is connected with the content of sulphur in the fuel and depends on the excess air coefficient in the piston engine. Figure 18 shows the dew-point temperature as the function of: sulphur content in the fuel, SO_2 conversion to SO_3, and the excess air coefficient in the engine. In inland power installations burning fuels with sulphur content higher than 2%, the recommended level of feed water temperature is t_{FW} > 140-145°C (Kehlhofer, 1991).

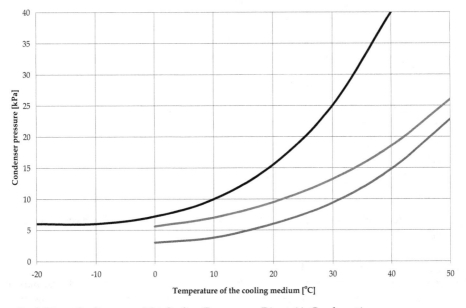

_____ Fresh Water Cooling   _____ Wet Cooling Tower   _____ Direct Air Condensation

Fig. 17. Condenser pressure as a function of temperature of the cooling medium

In marine propulsion (MAN, 1985) recommends that the feed water temperature should not be lower than 120°C when the sulphur content is higher than 2%. This is justified by the fact that the outer surface of the heater tubes on the exhaust gas side has the temperature higher by 8-15 °C than the feed water temperature, and that the materials used in those heaters reveal enhanced resistance to acid corrosion.

The exhaust gas temperature at the boiler outlet is assumed higher by 15-20°C than the feed water temperature, i.e. $t_{exh} > t_{FW} + (15 - 20°C)$.

Each ship burning heavy fuel in its power plant uses the mass flow rate m_{SS} of the saturated steam taken from the waste heat boiler for fuel pre-heating and all-ship purposes. According to the recommendations (MAN, 1985) the pressure of the steam used for these purposes should range between $p_{SS} = 7$-9 bar. This pressure is also assumed equal to the pressure in the boiler low-pressure circuit. The back temperature of the above steam flow in the heat box is within 50 – 60°C.

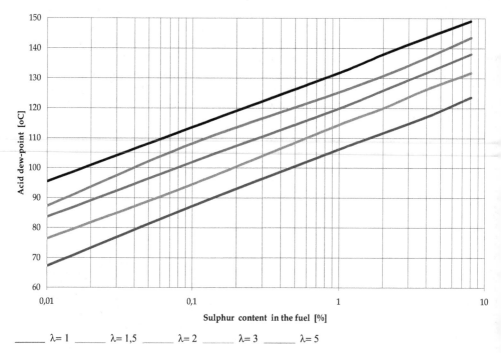

Fig. 18. Acid dew-point as a function of the sulphur content in the fuel and the excess air coefficient λ

4.2 Optimising the steam circuit

Optimisation of the steam system is to be done in such a way so as to reach the maximal possible utilisation of the heat contained in the exhaust gas. In this sense the optimisation is reduced to selecting the steam circuit parameters for which the steam turbine reaches the highest power. The area of search for optimal steam circuit parameters is to be narrowed to the sub-area where the earlier discussed limits imposed on the steam system are met. The use of the steam system with the waste heat boiler increases the power of the propulsion system within the entire range of the main engine load.

Adding a steam turbine to the Diesel engine system increases the power of the propulsion system by $\Delta N_{ST}/N_D = 6{,}5 - 7{,}5\%$ for main engine loads ranging from 90 to 100%. The power of the steam turbine for both examined variants of power turbine supply are comparable, and slightly higher power, by about 2-4%, is obtained by the steam turbine in the variant with series power turbine supply.

The analysis of the system with an additional exchanger utilising the low-temperature waste heat from the Diesel engine to heat the condensate from the condenser before the deaerator, Fig.16, shows that the steam turbine power increases by 7- 11% with respect to that of the steam turbine without this exchanger.

The requirements concerning the waste heat boiler refer to low loss of the exhaust gas flow (which reduces the final expansion pressure in the power turbine) and small temperature concentrations (pitch points) in the boiler evaporators. There is a remarkable impact of the sulphur content in the fuel on the permissible exhaust gas temperature and the lower feed water temperature limit. In the steam turbine circuit, a minimal number of exchangers should be used (optimally: none). The optimal parameters of this circuit also depend on the piston engine load.

5. Conclusions

It is possible to implement a combined system consisting of a Diesel engine as the leading engine, a power gas turbine, and a steam turbine circuit utilising the heat contained in the Diesel engine exhaust gas. Such systems can reveal thermodynamic efficiencies comparable with combined gas turbine circuits connected with steam turbines.

5.1 Power range of combined systems

Depending on the adopted variant and the main engine load, the use of the combined system makes it possible to increase the power of the power plant by 7 to 15 % with respect to the conventional power plant burning the same rate of fuel. Additional power is obtained by the system due to the recovery of the energy contained in the exhaust gas leaving the piston internal combustion engine. Thus the combined system decreases the specific fuel consumption by 6,4 - 12,8 % compared to the conventional power plant.

In the examined systems the power of the steam turbine is higher than that of the power turbine by 6-29 %, depending on the system variant and the main engine load.

5.2 Efficiency of combined systems

The use of the combined system for ship propulsion increases the efficiency of the propulsion system, and decreases the specific fuel consumption. Additionally, it increases the propulsion power without additional fuel consumption.

Like the power, the efficiency of the combined system increases with respect to the conventional power plant by 7 to 15% reaching the level of 53 - 56% for maximal power ranges. These efficiency levels are comparable with the combined systems based on the steam/gas turbines, Fig. 1. For partial loads the efficiency curves of the combined system with the Diesel engine are more flat than those for the combined turbine systems (smaller efficiency decrease following the load decrease) .

In the combined system the maximal efficiency is reached using particular system components:

- the piston internal combustion engine with the maximal efficiency;
- Turbocharger. The turbocharger with the maximal efficiency should be used as it provides opportunities for decreasing the exhaust gas enthalpy drop in the turbine in case of the series supply variant, or exhaust gas mass flow rate in case of the parallel supply variant, which in both cases results in higher power of the power turbine;
- Power turbine. High efficiency is required to increase its power;
- Steam turbine circuit. The requirement is to obtain the maximal power of the steam turbine from the heat delivered in the exhaust gas flowing through the boiler.

5.3 Ecology

Along with the thermodynamic profits, having the form of efficiency increase, and the economic gains, reducing the fuel consumption for the same power output of the propulsion system, the use of the combined system brings also ecological profits. A typical new-generation low-speed piston engine fed with heavy fuel oil with the sulphur content of 3% emits 17g/kWh NOx, 12g/kWh SOx and 600g/kWhCO$_2$ to the atmosphere. The use of the combined system reduces the emission of the noxious substances by, respectively, g/kWh NOx, g/kWh SOx and   g/kWhCO$_2$. The emission decreases by   % with respect to the standard engine, solely because of the increased system efficiency, without any additional installations.

Depending on the adopted solution, the combined power plant provides opportunities for reaching the assumed power of the propulsion system at a lower load of the main Diesel engine, at the same time also reducing the fuel consumption.

The article presents the thermodynamic analysis of the combined system consisting of the Diesel engine, the power gas turbine, and the steam turbine, without additional technical and economic analysis which will fully justify the application of this type of propulsion systems in power conversion systems.

6. Nomenclature

b_e	- specific fuel oil consumption
c_g, c_a	- specific heat of exhaust gas and air, respectively
i	- specific enthalpy
m	- mass flow rate
N	- power
p	- pressure
T,t	- temperature
Wu	- calorific value of fuel oil
η	- efficiency
κ_g, κ_a	- isentropic exponent of exhaust gas and air, respectively

Indices:

a	- air
bar	- barometric conditions
B	- Boiler
C	- Compressor
combi	- combined system
D	- Diesel engine
d	- supercharging
exh	- exhaust passage
f	- fuel
FW	- feet water
g	- exhaust gas
inlet	- inlet passage
k	- parameters in a condenser
o	- live steam, calculation point
PT	- Power turbine
ST	- Steam turbine
ss	- ship living purposes
T	- Turbine
TC	- Turbocharger
π	- compression ratio in a compressor, expansion ratio in a turbine

7. References

Dzida, M. (2009). On the possible increasing of efficiency of ship power plant with the system combined of marine diesel engine, gas turbine and steam turbine at the main engine - steam turbine mode of cooperation. *Polish Maritime Research,* Vol. 16, No.1(59), (2009), pp. 47-52, ISSN 1233-2585

Dzida, M. & Mucharski, J. (2009). On the possible increasing of efficiency of ship power plant with the system combined of marine diesel engine, gas turbine and steam turbine in case of main engine cooperation with the gas turbine fed in parallel and the steam turbine. *Polish Maritime Research,* Vol 16, No 2(60), pp. 40-44, ISSN 1233-2585

Dzida, M.; Girtler, J.; Dzida, S. (2009). On the possible increasing of efficiency of ship power plant with the system combined of marine diesel engine, gas turbine and steam turbine in case of main engine cooperation with the gas turbine fed in series and the steam turbine. *Polish Maritime Research,* Vol 16, No 3(61), pp. 26-31, ISSN 1233-2585

Kehlhofer, R. (1991). *Combined-Cycle Gas & Steam Turbine Power Plants,* The Fairmont Press, INC., ISBN 0-88173-076-9, USA

MAN B&M (October 1985). The MC Engine. Exhaust Gas Date. Waste Heat Recovery
    System. Total Economy, *MAN B&W Publication S.A.*, Danish

MAN Diesel & Turbo (2010). Stationary Engine. Programme 4th edition, *Branch of MAN
    Diesel & Turbo SE*, Germany, Available from www.mandieselturbo.com

Schrott, K. H. (1995). The New Generation of MAN B&W Turbochargers. *MAN B&W
    Publication S.A.*, No.236 5581E

3

State-of-Art of Transonic Axial Compressors

Roberto Biollo and Ernesto Benini
University of Padova
Italy

1. Introduction

Transonic axial flow compressors are today widely used in aircraft engines to obtain maximum pressure ratios per single-stage. High stage pressure ratios are important because they make it possible to reduce the engine weight and size and, therefore, investment and operational costs. Performance of transonic compressors has today reached a high level but engine manufacturers are oriented towards increasing it further. A small increment in efficiency, for instance, can result in huge savings in fuel costs and determine a key factor for product success. Another important target is the improvement of rotor stability towards near stall conditions, resulting in a wider working range.

Important analytical and experimental researches in the field of transonic compressors were carried out since 1960's (e.g. Chen et al., 1991; Epstein, 1977; Freeman & Cumpsty, 1992; König et al., 1996; Miller et al., 1961; Wennerstrom & Puterbaugh, 1984). A considerable contribution for the new developments and designs was the progress made in optical measurement techniques and computational methods, leading to a deeper understanding of the loss mechanisms of supersonic relative flow in compressors (e.g. Calvert & Stapleton, 1994; Hah & Reid, 1992; Ning & Xu, 2001; Puterbaugh et al., 1997; Strazisar, 1985; Weyer & Dunker, 1978). Fig. 1 shows the low pressure and high pressure compressors of the EJ200 engine as examples for highly loaded, high performance transonic rotors of an aero engine.

A closer look at the current trend in design parameters for axial flow transonic compressors shows that, especially in civil aircraft engines, the relative flow tip Mach number of the rotor is limited to maintain high efficiencies. A typical value for the rotor inlet relative flow at the tip is Mach ≈ 1.3. The continuous progress of aerodynamics has been focused to the increase in efficiency and pressure ratio and to the improvement in off-design behaviour at roughly the same level of the inlet relative Mach number. Today's high efficiency transonic axial flow compressors give a total pressure ratio in the order of 1.7-1.8, realized by combining high rotor speeds (tip speed in the order of 500 m/s) and high stage loadings ($2\Delta h/u^2$ in the order of 1.0). The rotor blade aspect ratio parameter showed a general trend towards lower values during past decades, with a current asymptotic value of 1.2 (Broichhausen & Ziegler, 2005).

The flow field that develops inside a transonic compressor rotor is extremely complex and presents many challenges to compressor designers, who have to deal with several and concurring flow features such as shock waves, intense secondary flows, shock/boundary layer interaction, etc., inducing energy losses and efficiency reduction (Calvert et al., 2003; Cumpsty, 1989; Denton & Xu, 1999; Law & Wadia, 1993; Sun et al., 2007). Interacting with secondary flows, shock waves concur in development of blockage (Suder, 1998), in corner stall separation (Hah & Loellbach, 1999; Weber et al., 2002), in upstream wakes destabilization (Estevadeordal et al., 2007; Prasad, 2003), and in many other negative flow phenomena. Particularly detrimental is the interaction with the tip clearance flow at the outer span of the rotor, where the compressor generally shows the higher entropy production (Bergner et al., 2005a; Chima, 1998; Copenhaver et al., 1996; Gerolymos & Vallet, 1999; Hofmann & Ballmann, 2002; Puterbaugh & Brendel, 1997; Suder & Celestina, 1996).

Fig. 1. Transonic LPC (left) and HPC (right) of the Eurofighter Typhoon engine EJ200 (Broichhausen & Ziegler, 2005)

As the compressor moves from peak to near-stall operating point, the blade loading increases and flow structures become stronger and unsteady. The tip leakage vortex can breakdown interacting with the passage shock wave, leading to not only a large blockage effect near the tip but also a self-sustained flow oscillation in the rotor passage. As a result, the blade torque, the low energy fluid flow due to the shock/tip leakage vortex interaction and the shock-induced flow separation on the blade suction surface fluctuate with time (Yamada et al., 2004).

Despite the presence of such flow unsteadiness, the compressor can still operate in a stable mode. Rotating stall arises when the loading is further increased, i.e. at a condition of lower mass flow rate. Two routes to rotating stall have been identified: long length-scale (modal) and short length-scale (spike) stall inception in axial compressors (Day, 1993). Modal stall inception is characterized by the relatively slow growth (over 10-40 rotor revolutions) of a small disturbance of long circumferential wavelength into a fully developed stall cell. Spike stall inception starts with the appearance of a large amplitude short length-scale (two to three rotor blade passages) disturbance at the rotor tip, the so-called spike, which grows into a fully developed rotating stall cell within few rotor revolutions.

The following paragraphs give a summary of the possible techniques for limiting the negative impacts of the above reported compressor flow features in aircraft gas turbine engines.

2. Blade profiles studies

For relative inlet Mach numbers in the order of 1.3 and higher the most important design intent is to reduce the Mach number in front of the passage shock. This is of primary importance due to the strongly rising pressure losses with increasing pre-shock Mach number, and because of the increasing pressure losses due to the shock/boundary layer interaction or shock-induced separation. The reduction of the pre-shock Mach number can be achieved by zero or even negative curvature in the front part of the blade suction side and by a resulting pre-compression shock system reducing the Mach number upstream of the final strong passage shock.

Besides inducing energy losses, the presence of shock waves makes transonic compressors particularly sensitive to variations in blade section design. An investigation of cascade throat area, internal contraction, and trailing edge effective camber on compressor performance showed that small changes in meanline angles, and consequently in the airfoil shape and passage area ratios, significantly affect the performance of transonic blade rows (Wadia & Copenhaver, 1996).

One of the most important airfoil design parameter affecting the aerodynamics of transonic bladings is the chordwise location of maximum thickness. An experimental and numerical evaluation of two versions of a low aspect ratio transonic rotor having the location of the tip blade section maximum thickness at 55% and 40% chord length respectively, showed that the more aft position of maximum thickness is preferred for the best high speed performance, keeping the edge and maximum thickness values the same (Wadia & Law, 1993). The better performance was associated with the lower shock front losses with the finer section that results when the location of the maximum thickness is moved aft. The existence of an optimum maximum thickness location at 55% to 60% chord length for such rotor was hypothesized. Similar results can be found in a recent work (Chen et al., 2007) describing an optimization methodology for the aerodynamic design of turbomachinery applied to a transonic compressor bladings and showing how the thermal loss coefficient decreases with increasing the maximum thickness location for all the sections from hub to tip.

Not only the position of maximum thickness but also the airfoil thickness has been showed to have a significant impact on the aerodynamic behaviour of transonic compressor rotors, as observed in an investigation on surface roughness and airfoil thickness effects (Suder et al., 1995). In this work, a 0.025 mm thick smooth coating was applied to the pressure and suction surface of the rotor blades, increasing the leading edge thickness by 10% at the hub and 20% at the tip. The smooth coating surface finish was comparable to the bare metal blade surface finish; therefore the coating did not increase roughness over the blade, except at the leading edge where roughness increased due to particle impact damage. It resulted in a 4% loss in pressure ratio across the rotor at an operating point near design mass flow, with the largest degradation in pressure rise over the outer half of the blade span. When assessed at a constant pressure ratio, the adiabatic efficiency degradation at design speed was in the order of 3-6 points.

The recent development of optimization tools coupled with accurate CFD codes has improved the turbomachinery design process significantly, making it faster and more efficient. The application to the blade section design, with a quasi three-dimensional and more recently with a fully three-dimensional approach, can lead to optimal blade geometries in terms of aerodynamic performance at both design and off-design operating conditions. Such a design process is particularly successful in the field of transonic compressors, where performance is highly sensitive to little changes in airfoil design.

Fig. 2 shows the blade deformation obtained in a quasi 3-D numerical optimization process of a transonic compressor blade section along with the relative Mach number contours before and after the optimization (Burguburu et al., 2004). As shown, no modifications of the inlet flow field occurred after optimization but the flow field structure in the duct is clearly different. The negative curvature of the blade upstream of the shock led to the reduction of the upstream relative Mach number from 1.4 to 1.2. With this curvature change, the velocity slowdown is better driven. Instead of creating a normal shock, the new shape created two low intensity shocks. The new blade gave an efficiency increment of 1.75 points at design condition, without changing the choking mass flow. A large part of the efficiency improvement at the design condition remained at off-design conditions.

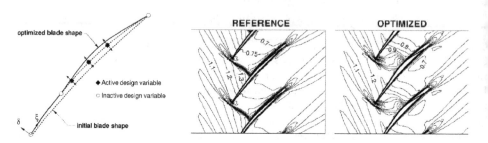

Fig. 2. Blade deformation (left) and relative Mach number contours (right) before and after optimization (Burguburu et al., 2004)

Fig. 3 is related to a both aerodynamic and structural optimization of the well-known transonic compressor rotor 67 (Strazisar et al., 1989), where the aerodynamic objective aimed at maximizing the total pressure ratio whereas the structural objective was to minimize the blade weight, with the constraint that the new design had comparable mass flow rate as the baseline design (Lian & Liou, 2005). The optimization was carried out at the design operating point. Geometric modifications regarded the mean camber line (with the leading and trailing edge points fixed) and thickness distribution of four airfoil profiles (hub, 31% span, 62% span, and tip), linearly interpolated to obtained the new 3-D blade. The chord distribution along the span and the meridional contours of hub, casing, sweep, and lean were maintained.

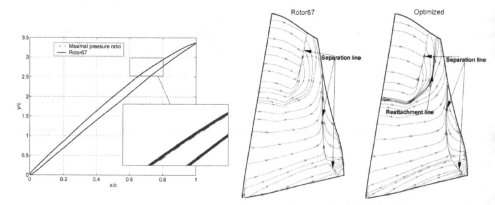

Fig. 3. Blade section at 90% span (left) and streamlines close to the blade suction side (right) before and after the optimization (modified from Lian & Liou, 2005)

At 10% and 50% span (not shown here), the optimization gave a larger camber but lower thickness than the baseline design. The thinner airfoils contributed to reduce the weight of the new design. The calculated difference in the pressure distribution was rather small. At 90% span (see Fig. 3), the new design had a slightly smaller camber and thinner airfoil than the baseline. Nevertheless, the calculated pressure difference was rather large, indicating again that transonic flow is highly sensitive to the profile shape change. One noticeable impact was also in the shock position. The new design showed a more forward passage shock than the baseline.

Such optimized blade gave a decrease of 5.4% in weight and an improvement of 1.8% in the total pressure ratio. The lighter weight came from the thinner blade shape. The higher total pressure ratio was mainly attributed to a reduced separation zone after the shock at the outer span. In Fig. 3, the separation zones are characterized by streamlines going towards the separation lines, whereas reattachment lines look like flow is going away from the separation lines. Compared with the baseline design, downstream of the shock the new design gave a smaller separation zone, which was partially responsible for its higher total pressure ratio.

Fig. 4 is again related to the redesign of rotor 67 using an optimization tool based on evolutionary algorithms (Oyama et al., 2004). Note the particular new design, an improbable design using manual techniques. The optimization gave rise to a double-hump blade shape, especially obvious on the pressure side.

In such new design, the flow acceleration near the leading edge at 33% span diminished because of the decrease of the incidence angle. In addition, at the 90% span, the shock on the suction side moved aft and was weakened considerably because of the aft movement of the maximum camber position. This new blade showed an overall adiabatic efficiency of 2% higher than the baseline blade over the entire operating range for the design speed.

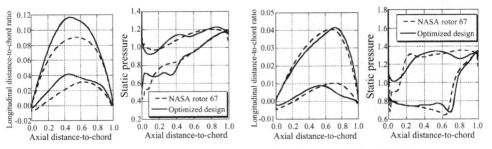

Fig. 4. Comparison between the optimized and baseline design at 33% and 90% span (Oyama et al., 2004)

3. Three-dimensional shaped bladings

The preceding paragraph has shown that a certain maturity in transonic compressors has been reached regarding the general airfoil aerodesign. But the flow field in a compressor is not only influenced by the two-dimensional airfoil geometry. The three-dimensional shape of the blade is also of great importance, especially in transonic compressor rotors where an optimization of shock structure and its interference with secondary flows is required. Many experimental and numerical works can be found in the literature on the design and analysis of three-dimensional shaped transonic bladings (e.g. Copenhaver et al., 1996; Hah et al.,

2004; Puterbaugh et al., 1997). Fig. 5 shows two examples of non-conventional rotors (Rotor2 and Rotor3) derived from the baseline Rotor1 which is conventionally radially-stacked, all developed by TU Darmstadt and MTU Aero Engines. As far as their performance is concerned, Rotor2 gave no real improvement in efficiency and total pressure ratio with respect to the baseline configuration (Blaha et al., 2000; Kablitz et al., 2003a). Rotor3, instead, gave higher performance at design speed (1.5% peak efficiency increment) along with a significantly wider operating range (Passrucker et al., 2003). Information on the favourable impact of Rotor3 blade design on internal transonic flow field is available in the open literature (Bergner et al., 2005b; Kablitz et al., 2003b).

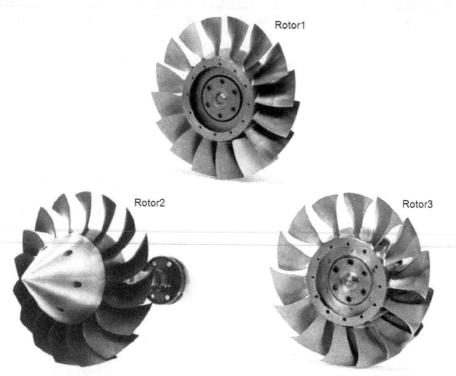

Fig. 5. Transonic compressor test rotors – TU Darmstadt and MTU Aero Engines (Broichhausen & Ziegler, 2005; Passrucker et al., 2003)

A numerical investigation on the aerodynamics of 3-D shaped blades in transonic compressor rotors showed the possibility to have better stall margin with forward sweep (upstream movement of blade sections along the local chord direction, especially at outer span region), maintaining a high efficiency over a wider range (Denton, 2002; Denton & Xu, 2002). This seems to be a general point of view, as confirmed by the following researches.

Numerical and experimental analyses carried out to evaluate the performance of a conventional unswept rotor, a forward swept rotor and an aft swept rotor showed that the forward swept rotor had a higher peak efficiency and a substantially larger stall margin than the baseline unswept rotor, and that the aft swept rotor had a similar peak efficiency with a significantly smaller stall margin (Hah et al., 1998). Detailed analyses of the measured and calculated flow fields indicated that two mechanisms were primarily responsible for the differences in aerodynamic performance among these rotors. The first mechanism was a change in the radial shape of the passage shock near the casing by the endwall effect, and the second was the radial migration of low momentum fluid to the blade tip region. Similar results were obtained in a parallel investigation which identified the reduced shock/boundary layer interaction, resulting from reduced axial flow diffusion and less accumulation of centrifuged blade surface boundary layer at the tip, as the prime contributor to the enhanced performance with forward sweep (Wadia et al., 1998).

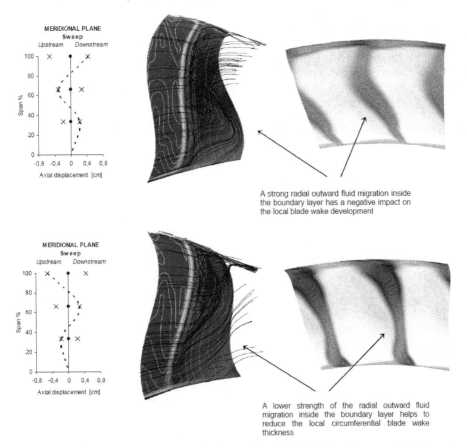

A strong radial outward fluid migration inside the boundary layer has a negative impact on the local blade wake development

A lower strength of the radial outward fluid migration inside the boundary layer helps to reduce the local circumferential blade wake thickness

Fig. 6. Blade axial curvature impact on shock, suction side boundary layer and blade wake development (Biollo & Benini, 2008a)

A recent numerical work gave another point of view on the impact of blade curvature in transonic compressor rotors, showing how the movement of blade sections in the axial direction can influence the internal flow field (Benini & Biollo, 2007; Biollo & Benini, 2008a). Such work showed that the axial blade curvature can help to influence the shock shape in the meridional plane, inducing the shock to assume the meridional curvature of the blade leading edge (Fig. 6). In addition, a considerable impact on the radial outward migration of fluid particles which takes place inside the blade suction side boundary layer after the interaction with the shock has been confirmed. The code predicted a reduction of the strength of such flow feature when the blade is curved downstream and an increment when the blade is curved upstream. Such flow phenomenon is harmful because obstructs the boundary layer development in the streamwise direction, leading to a thickening of blade wakes. A reduction of its strength helped to reduce the entropy generation and the aerodynamic losses associated with the blade wake development. The possibility to increase the peak efficiency of 0.8% at design speed using a proper downstream blade curvature has been showed for the high loaded transonic compressor rotor 37. Details on rotor 37 can be found in the open literature (Reid & Moore, 1978).

The same research group investigated the aerodynamic effects induced by several tangential blade curvatures on the same rotor. It was observed that, when the curvature is applied towards the direction of rotor rotation, the blade-to-blade shock tends to move more downstream, becoming more oblique to the incoming flow. This reduced the aerodynamic shock losses and entropy generation, showing in some cases a peak efficiency increment of over 1% at design speed (Benini & Biollo, 2008). Similar results were previously obtained using a numerical optimization algorithm (Ahn & Kim, 2002). Fig. 7 shows the predicted impact of the optimized design of rotor 37 on the blade-to-blade Mach number.

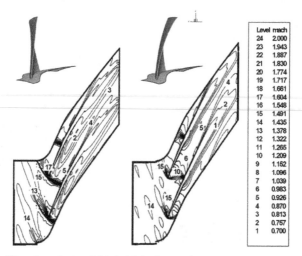

Level	mach
24	2.000
23	1.943
22	1.887
21	1.830
20	1.774
19	1.717
18	1.661
17	1.604
16	1.548
15	1.491
14	1.435
13	1.378
12	1.322
11	1.265
10	1.209
9	1.152
8	1.096
7	1.039
6	0.983
5	0.926
4	0.870
3	0.813
2	0.757
1	0.700

Fig. 7. Baseline (left) and optimized (right) Mach number distributions at 90% span (modified from Ahn & Kim, 2002)

Higher performance can be achieved using a proper combination of two orthogonal blade curvatures, i.e. the use of a blade curved both axially and tangentially, as well as swept and leaned at the same time. Peak efficiency increments from 1% to 1.5% were numerically observed using a blade prevalently curved towards the direction of rotor rotation and slightly backward inclined (Biollo & Benini, 2008b; Jang et al., 2006; Yi et al., 2006).

4. Casing treatments

Hollow structures in the casing to improve the tip endwall flow field of axial flow compressors are commonly referred to as casing treatments. Fig. 8 shows some examples of casing treatments investigated in the 1970's. The interaction of the main flow with the flow circulating in these cavities seems to have a positive impact on rotor stability. However, early studies did not reach a detailed understanding of the phenomenon, since experimental investigations were too expensive and only few configurations could be tested. Only in the past fifteen years numerical simulations made it possible to investigate a larger number of casing treatment solutions and their effects on different compressors. Many researches were carried out on transonic compressor rotors and the potential of this kind of passive devices was revealed: a proper treatment can not only widen the stable working range of a transonic compressor rotor, but also improve its efficiency.

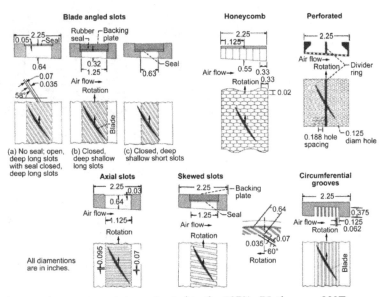

Fig. 8. Various casing treatments investigated in the 1970's (Hathaway, 2007)

4.1 Circumferential groove-type treatments

Recently, the influence of circumferential grooves on the tip flow field of an axial single-stage transonic compressor has been examined both experimentally and numerically (Fig. 9). The compressor stage provided a strongly increased stall margin (56.1%), with only small penalties in efficiency when the casing treatment was applied. Flow analyses showed that at near stall conditions with the smooth casing, the induced vortex originating from the tip clearance flow crossing the tip gap along 20-50% chord length, hits the front part of the adjacent blade, indicating the possibility of a spill forward of low momentum fluid into the next passage, a flow feature considered a trigger for the onset of rotating stall. With the casing treatment applied, the vortex trajectory remained instead aligned to the blade's suction side.

Disadvantages of casing treatments like these are the space they need and the weight increase of the compressor casing. So it is a goal to maintain the positive effects (increased surge pressure ratio in combination with high efficiency) while at the same time reducing the geometric volume of the device. On this regard, an experimental and numerical investigation on the first rotor of a two-stage compressor showed that grooves with a much smaller depth than conventional designs are similarly effective in increasing the stall margin (Rabe & Hah, 2002). The same work also showed that two shallow grooves placed near the leading edge are better than five deep or shallow grooves all over the blade tip. Fewer shallower grooves clearly help to reduce the weight, fabrication costs, and loss generation associated with such a casing treatment.

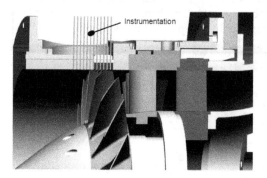

Fig. 9. Cross section of a transonic compressor stage with circumferential grooves (modified from Müller et al., 2008)

Other possible groove-type casing treatment solutions are presented in Figs. 10 and 11. Fig. 10 shows a single extended casing circumferential groove all over the blade tip section. It has been numerically shown that such a casing treatment provides a means for fluid to exit the flow path where the blade loading is severe, migrate circumferentially, and re-enter the flow path at a location where the pressure is more moderate. This can lead to stability improvement since the flow relocation helps to relieve the locally severe blade loading. Using this device, the authors showed the possibility to improve both the efficiency and the stall margin.

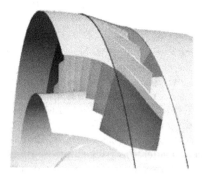

Fig. 10. Single extended casing circumferential groove (Beheshti et al., 2004)

Fig. 11 is related to a numerical investigation of casing contouring effects on flow instability. While the "Type B1" solution gave no improvements in stall margin, the "Type C2" solution extended the calculated stable operating range from 94% to about 92.5% normalized mass flow. With respect to the baseline smooth casing configuration, the successful contouring induced a smaller inflow angle near the leading edge, i.e. a lower incidence, delaying rotating stall inception.

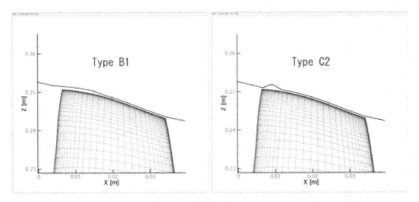

Fig. 11. Endwall casing contouring (Ito et al., 2007)

4.2 Slot-type treatments

Another way to treat the casing with the aim to improve the rotor performance and stability is exemplified in Fig. 12. Here the casing wall is circumferentially treated with a discrete number of axial rectangular slots over the blade tip section.

A similar slot-type casing treatment consisting of four identical axial slots per blade passage and having an open area of 50% in the circumferential direction has been also proposed. (Wilke and Kau, 2004). In this case, the slots are parallel to the rotation axis and inclined by 45° against the meridional plane in the direction of rotor rotation. The slot shape is designed as a semi-circle. Two configurations have been numerically tested. In configuration 1, the position of the slots is centered above the rotor blade tip reaching from 7.5% to 92.5% chord length. In configuration 2, the slots are moved upstream so that only 25% chord length remains covered by the casing treatment.

Fig. 12. EJ200 LPC with axial slots casing treatment (Broichhausen & Ziegler, 2005)

For both configurations, simulations showed a significant increment in flow stability compared to the solid wall, the stalling mass flow passed from 0.95% to 0.75% the design mass flow at the design speed. It was observed that the stabilizing effects are based on the positive impact of casing treatment on the tip clearance flow and its resulting vortex. Configuration 1 led to a massive destruction of the tip leakage vortex, whereas configuration 2 weakened the rolling-up of the tip clearance flow. Configuration 2, gave also a positive impact on the overall efficiency.

5. Air injection and bleeding

One of the first flow control ideas to receive a considerable attention in gas turbine applications is the flow injection and bleeding concept. Air injection near the blade tips has proved to increase compressor stall margin, leading to higher engine operability. Fig. 13 shows a set of successful 12 non-intrusive prototype injectors recently installed in the casing of a transonic compressor test rotor. On the other hand, aspiration (or bleeding) can be used to delay blade separation, which limits the stage work and therefore increases the required number of stages. Doubling the work per stage using aspiration results in a dramatic reduction in the number of stages, though not necessarily an exactly proportional reduction in compressor weight due to the added complexity of blades and added plumbing. For fighter applications this technology looks very attractive due to its potential to improve the thrust/weight ratio of the engine.

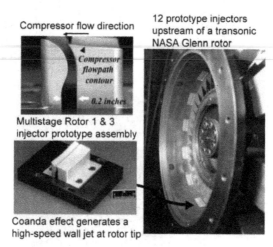

Fig. 13. Discrete tip injection (Hathaway, 2007)

5.1 Air injection

Measurements and simulations for active discrete tip injection have been presented for a range of steady injection rates and distributions of injectors around the annulus of a tip-critical transonic compressor rotor operating in isolation (Suder et al., 2001). In such case, casing-mounted injectors were located at 200% tip axial chord upstream of the rotor. Each injector penetrated 6% span from the casing into the flow field. The simulations indicated that tip injection increases stability by unloading the rotor tip. Tip injection decreases incidence and blade loading at the tip, allowing increased loading at lower blade spans before the blade stalls. With tip injection present, the blade stalls when the loading at the tip reaches the level equal to that for which the blade stalls with no injection. The experiments showed that stability enhancement is related to the mass averaged axial velocity at the tip. For the tested rotor, experimental results demonstrated that at 70% speed the stalling flow coefficient can be reduced by 30% using an injected mass flow equivalent to 1% of the annulus flow. At design speed, the stalling flow coefficient was reduced by 6% using an injected mass flow equivalent to 2% of the annulus flow. Tip injection has also been demonstrated as an effective tool for recovering a compressor from fully developed stall.

In a self-induced (passive) solution, steady discrete tip injection using casing recirculation has been simulated in a single transonic fan rotor (Hathaway, 2002). The idea was to bleed pressurized fluid from downstream of the rotor and properly inject it upstream. Simulations were carried out assuming both a clear and distorted inlet flow. The distortion was circumferentially applied near the casing endwall fixing a lower inlet total pressure. The recirculation model gave 125% range extension without inlet distortion and 225% range extension with inlet distortion, without significant impact on overall efficiency.

Such a solution has been successively tested on a single-stage transonic compressor (Strazisar et al., 2004). Results clearly indicated that recirculation extends the stable operating range of that stage to lower mass flows than that can be achieved without recirculation. With the recirculation activated, the positive change in stalling flow coefficient was 6% at 70% of design speed and 2% at design speed, with a total injected flow of 0.9% of the annulus flow at both operating speeds. In the same work, the potential for using endwall recirculation to increase the stability of transonic highly loaded multistage compressors was demonstrated through results from a rig test of simulated recirculation driving both a steady injected flow and an unsteady injected flow. Unsteady injection increased stability more than steady injection and was capable of changing the unsteady near stall dynamics of the multistage compressor.

5.2 Bleeding

Fig. 14 shows the sketch of a transonic aspirated stage experimentally tested and numerically investigated to demonstrate the application of boundary layer aspiration for increasing the stage work (Schuler et al., 2005). The stage was designed to produce a pressure ratio of 1.6 at a tip speed of 750 ft/s resulting in a stage work coefficient of 0.88. The primary aspiration requirement for the stage was a bleed fraction 0.5% of the inlet mass flow on the rotor and stator suction surfaces. Additional aspiration totalling 2.8% was also used at shock impingement locations and other locations on the hub and casing walls. The stage achieved a peak pressure ratio of 1.58 and through flow efficiency of 90% at the design point. The rotor showed an extremely high efficiency of 97% for a transonic rotor, partially attributed to aspiration and partially to the elimination of the tip clearance flow due to the tip shroud. Aspiration was also effective in maintaining stage performance at off-design conditions. The experimental data showed unstalled stage performance at least 83% of the design mass flow.

The possibility of a very high pressure ratio per single-stage using aspiration has been demonstrated for the fan of Fig. 15. The fan stage has been designed to achieve a pressure ratio of 3.4 at 1500 ft/s. The low energy viscous flow was aspirated from diffusion-limiting locations on the blades and flow path surfaces. Experimental results gave a stage pressure ratio exceeding 3 at design speed, with an aspiration flow fraction of 3.5% of the stage inlet mass flow. CFD simulations showed that aspiration fixes the passage shock position, particularly in the tip region, maintaining good aerodynamic behaviour at off-design operating points.

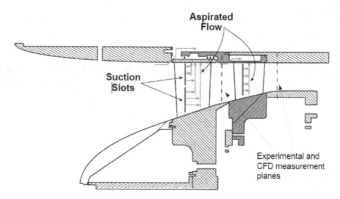

Fig. 14. A tested suction configuration (Schuler et al., 2002)

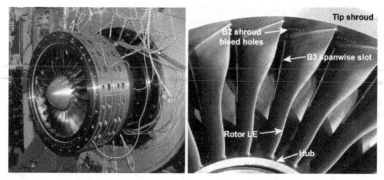

Fig. 15. Test package (left) and rotor aspiration scheme (right) (Merchant et al., 2005)

In a previous work (Dang & Van Rooij, 2003), it was demonstrated the possibility to reduce the amount of aspiration without compromise its benefits. When used as an add-on to an existing design, aspiration can require excessive amounts of suction, whereas with a blade/ aspiration scheme coupled design a significantly lower amount of suction can be needed. A baseline supersonic rotor with 56 blades was used in that work as the starting point. The blade count in that rotor was then reduced to 42 blades, and it was found that 2% of flow suction was needed to pull the shock back into the blade passage for the required back pressure. The aspiration zone was located approximately behind the shock on the suction surface from hub to 95% span. With 42 blades, inspection of the flow field showed that the peak Mach number and loading were significantly higher than in the baseline blade case, resulting in increased shock losses, thickening of the blade suction surface boundary layer, and a large region of low momentum fluid in the tip clearance region.

A new pressure loading shape was developed to mitigate the situation. The new design was shown to have better efficiency potential and a cleaner casing endwall flow using only 0.3%

flow suction. It was also shown that the constant speed throttling characteristic of the new design can be tailored by using varying amounts of suction (up to 2%) to hold the passage shock stationary thereby making it possible to satisfy multiple operating conditions with minimal aerodynamic penalty.

6. Conclusions

Transonic compressors are the state-of-art in the compression system of today's civil and military aero engines. Their capability to provide high pressure ratios maintaining high efficiencies made them preferable to both subsonic (lower pressure ratio) and supersonic (lower efficiency) solutions.

The researches of last decades have greatly contributed to their success. The progress made in optical measurement techniques and the development of computational methods led to a deep understanding of the loss mechanisms associated with their inherent supersonic relative flow, making it possible significant aerodynamic improvements and industrial diffusion.

Nowadays there is still space for further radical improvements and new solutions will be developed in the next future, like highly effective stall/surge control systems and higher pressure ratio configurations, or even new concepts based on new advanced working principles.

7. References

Ahn, C.-S. & Kim, K.-Y. (2002). Aerodynamic Design Optimization of an Axial Flow Compressor Rotor, *Proceedings of ASME Turbo Expo 2002*, GT-2002-30445

Beheshti, B. H.; Teixeira, J. A.; Ivey, P. C.; Ghorbanian, K. & Farhanieh, B. (2004). Parametric Study of Tip Clearance - Casing Treatment on Performance and Stability of a Transonic Axial Compressor, *Proceedings of ASME Turbo Expo 2004*, GT2004-53390

Benini, E. & Biollo, R. (2007). Aerodynamics of Swept and Leaned Transonic Compressor-Rotors, *Applied Energy*, Vol. 84, No. 10, (October 2007), pp. 1012-1027, ISSN 0306-2619

Benini, E. & Biollo, R. (2008). Effect of Forward and Aft Lean on the Performance of a Transonic Compressor Rotor, *International Journal of Turbo and Jet Engines*, Vol. 25, No. 1, (2008), pp. 13-26, ISSN 0334-0082

Bergner, J.; Hennecke, D. K. & Hah, C. (2005a). Tip-Clearance Variations of an Axial High-Speed Single-Stage Transonic Compressor, *Proceedings of 17th Symposium on Air-Breathing Engines*, ISABE-2005-1096

Bergner, J.; Kablitz, S.; Hennecke, D. K.; Passrucker, H. & Steinhardt, E. (2005b). Influence of Sweep on the 3D Shock Structure in an Axial Transonic Compressor, *Proceedings of ASME Turbo Expo 2005*, GT2005-68835

Biollo, R. & Benini, E. (2008a). Impact of Sweep and Lean on the Aerodynamic Behaviour of Transonic Compressor Rotors, *Proceedings of The Future of Gas Turbine Technology - 4th International Conference*, IGTC08_P17

Biollo, R. & Benini, E. (2008b). Aerodynamic Behaviour of a Novel Three-Dimensional Shaped Transonic Compressor Rotor Blade, *Proceedings of ASME Turbo Expo 2008*, GT2008-51397

Blaha, C.; Kablitz, S.; Hennecke, D. K.; Schmidt-Eisenlohr, U.; Pirker, K. & Haselhoff, S. (2000). Numerical Investigation of the Flow in an Aft-Swept Transonic Compressor Rotor, *Proceedings of ASME Turbo Expo 2000*, 2000-GT-0490

Broichhausen, K. D. & Ziegler, K. U. (2005). Supersonic and Transonic Compressors: Past, Status and Technology Trends, *Proceedings of ASME Turbo Expo 2005*, GT2005-69067

Burguburu, S.; Toussaint, C.; Bonhomme, C. & Leroy, G. (2004). Numerical Optimization of Turbomachinery Bladings, *ASME Journal of Turbomachinery*, Vol. 126, No. 1, (January 2004), pp. 91-100, ISSN 0889-504X

Calvert, W. J.; Emmerson, P. R. & Moore J. M. (2003). Design, Test and Analysis of a High-Pressure-Ratio Transonic Fan, *Proceedings of ASME Turbo Expo 2003*, GT2003-38302

Calvert, W. J. & Stapleton, A.W. (1994). Detailed Flow Measurements and Predictions for a Three-Stage Transonic Fan, *ASME Journal of Turbomachinery*, Vol.116, No.2, (April 1994), pp. 298-305, ISSN 0889-504X

Chen, G. T.; Greitzer, E. M.; Tan, C. S. & Marble, F. E. (1991). Similarity Analysis of Compressor Tip Clearance Flow Structure, *ASME Journal of Turbomachinery*, Vol. 113, No. 2, (April 1991), pp. 260-229, ISSN 0889-504X

Chen, N.; Zhang, H.; Xu, Y. & Huang, W. (2007). Blade Parameterization and Aerodynamic Design Optimization for a 3D Transonic Compressor Rotor, *Journal of Thermal Science*, Vol. 16, No. 2, (May 2007), pp. 105-114, ISSN 1003-2169

Chima, R. V. (1998). Calculation of Tip Clearance Effects in a Transonic Compressor Rotor, *ASME Journal of Turbomachinery*, Vol. 120, No. 1, (January 1998), pp. 131-140, ISSN 0889-504X

Copenhaver, W. W.; Mayhew, E. R.; Hah, C. & Wadia, A. R. (1996). The Effect of Tip Clearance on a Swept Transonic Compressor Rotor, *ASME Journal of Turbomachinery*, Vol. 118, No. 2, (April 1996), pp. 230-229, ISSN 0889-504X

Cumpsty, N. A. (1989). Compressor Aerodynamics, Longman Group UK Limited 1989

Dang, T.Q. & Van Rooij, M. (2003). Design of Aspirated Compressor Blades Using Three-Dimensional Inverse Method, NASA Report No. TM-2003-212212

Day, I. J. (1993). Stall Inception in Axial Flow Compressors, *ASME Journal of Turbomachinery*, Vol. 115, No. 1, (January 1993), pp. 1-9, ISSN 0889-504X

Denton, J. D. (2002). The Effects of Lean and Sweep on Transonic Fan Performance: A Computational Study, *Task Quarterly*, Vol. 6, No. 1, (2002), pp. 7-23

Denton, J. D. & Xu, L. (1999). The Exploitation of Three-Dimensional Flow in Turbomachinery Design, *Proc Instn Mech Engrs, Part C: Journal of Mechanical Engineering Science*, Vol. 213, No. 2, (1999), pp. 125-137, ISSN 0954-4062

Denton, J. D. & Xu, L. (2002). The Effects of Lean and Sweep on Transonic Fan Performance, *Proceedings of ASME Turbo Expo 2002*, GT-2002-30327

Epstein, A. H. (1977). Quantitative Density Visualization in a Transonic Compressor Rotor, *ASME Journal of Engineering for Power*, Vol.99, No.3, (July 1977), pp. 460-475, ISSN 0022-0825

Estevadeordal, J.; Gorrell, S. E. & Copenhaver, W. W. (2007). PIV Study of Wake-Rotor Interactions in a Transonic Compressor at Various Operating Conditions, *Journal of Propulsion and Power*, Vol. 23, No. 1, (Jan-Feb 2007), ISSN 0748-4658

Freeman, C. & Cumpsty, N. A. (1992). Method for the Prediction of Supersonic Compressor Blade Performance, *Journal of Propulsion and Power*, Vol. 8, No. 1, (Jan-Feb 1992), pp. 199-208, ISSN 0748-4658

Gerolymos, G. A. & Vallet, I. (1999). Tip-Clearance and Secondary Flows in a Transonic Compressor Rotor, *ASME Journal of Turbomachinery*, Vol. 121, No. 4, (October 1999), pp. 751-762, ISSN 0889-504X

Hah, C . & Loellbach, J. (1999). Development of Hub Corner Stall and Its Influence on the Performance of Axial Compressor Blade Rows, *ASME Journal of Turbomachinery*, Vol. 121, No. 1, (January 1999), pp. 67-77, ISSN 0889-504X

Hah, C.; Puterbaugh, S. L. & Wadia, A. R. (1998). Control of Shock Structure and Secondary Flow Field Inside Transonic Compressor Rotors Through Aerodynamic Sweep, *Proceedings of the 1998 International Gas Turbine & Aeroengine Congress & Exibition*, ASME Paper 98-GT-561

Hah, C.; Rabe, D. C. & Wadia, A. R. (2004). Role of Tip-Leakage Vortices and Passage Shock in Stall Inception in a Swept Transonic Compressor Rotor, *Proceedings of ASME Turbo Expo 2004*, GT2004-53867

Hah, C. & Reid, L. (1992). A Viscous Flow Study of Shock-Boundary Layer Interaction, Radial Transport, and Wake Development in a Transonic Compressor, *ASME Journal of Turbomachinery*, Vol. 114, No. 3, (July 1992), pp. 538-547, ISSN 0889-504X

Hathaway, M. D. (2002). Self-Recirculating Casing Treatment Concept for Enhanced Compressor Performance, *Proceedings of ASME Turbo Expo 2002*, GT-2002-30368

Hathaway, M. D. (2007). Passive Endwall Treatments for Enhancing Stability, NASA Report No. TM-2007-214409

Hofmann, W. & Ballmann, J. (2002). Tip Clearance Vortex Development and Shock-Vortex-Interaction in a Transonic Axial Compressor Rotor, *Proceedings of 40th AIAA Aerospace Sciences Meeting and Exhibit*, AIAA 2002-0083

Ito, Y.; Watanabe, T. & Himeno, T. (2007). Effect of Endwall Contouring on Flow Instability of Transonic Compressor, *Proceedings of the International Gas Turbine Congress 2007 Tokyo*, IGTC2007 Tokyo TS-042

Jang, C.-M.; Samad, A. & Kim, K.-Y. (2006). Optimal Design of Swept, Leaned and Skewed Blades in a Transonic Axial Compressor, *Proceedings of ASME Turbo Expo 2006*, GT2006-90384

Kablitz, S.; Bergner, J.; Hennecke, K.; Beversdorff, M. & Schodl, R. (2003a). Darmstadt Rotor No. 2, III: Experimental Analysis of an Aft-Swept Axial Transonic Compressor Stage, *International Journal of Rotating Machinery*, Vol. 9, No. 6, (2003), pp. 393-402, ISSN 1023-621X

Kablitz, S.; Passrucker, H.; Hennecke, D. K. & Engber, M. (2003b). Experimental Analysis of the Influence of Sweep on Tip Leakage Vortex Structure of an Axial Transonic Compressor Stage, *Proceedings of 16th Symposium on Air-Breathing Engines*, ISABE-2003-1226

König, W. M.; Hennecke, D. K. & Fottner L. (1996). Improved Blade Profile Loss and Deviation Angle Models for Advanced Transonic Compressor Bladings: Part II-A Model for Supersonic Flow, *ASME Journal of Turbomachinery*, Vol. 118, No. 1, (January 1996), pp. 81-87, ISSN 0889-504X

Law, C. H. and Wadia, A. R. (1993). Low Aspect Ratio Transonic Rotors: Part 1-Baseline Design and Performance, *ASME Journal of Turbomachinery*, Vol. 115, No. 2, (April 1993), pp. 218-225, ISSN 0889-504X

Lian, Y. & Liou, M.-S. (2005). Multi-Objective Optimization of Transonic Compressor Blade Using Evolutionary Algorithm, *Journal of Propulsion and Power*, Vol. 21, No. 6, (Nov-Dec 2005), pp. 979-987, ISSN 0748-4658

Merchant, A.; Kerrebrock, J. L.; Adamczyk, J. J. & Braunscheidel, E. (2005). Experimental Investigation of a High Pressure Ratio Aspirated Fan Stage, *ASME Journal of Turbomachinery*, Vol. 127, No. 1, (January 2005), pp. 43-51, ISSN 0889-504X

Miller, G. R.; Lewis, Jr., G. W. & Hartmann, M. J. (1961). Shock Losses in Transonic Compressor Blade Rows, *ASME Journal of Engineering for Power*, Vol. 83, No. 3, (July 1961), pp. 235-242, ISSN 0022-0825

Müller, M. W.; Biela, C.; Schiffer H.-P. & Hah, C. (2008). Interaction of Rotor and Casing Treatment Flow in an Axial Single-Stage Transonic Compressor With Circumferential Grooves, *Proceedings of ASME Turbo Expo 2008*, GT2008-50135

Ning, F. & Xu, L. (2001). Numerical Investigation of Transonic Compressor Rotor Flow Using an Implicit 3D Flow Solver With One-Equation Spalart-Allmaras Turbulence Model, *Proceedings of ASME Turbo Expo 2001*, 2001-GT-0359

Oyama, A.; Liou, M.-S. & Obayashi, S. (2004). Transonic Axial-Flow Blade Optimization: Evolutionary Algorithms/Three-Dimensional Navier-Stokes Solver, *Journal of Propulsion and Power*, Vol. 20, No. 4, (Jul-Aug 2004), pp. 612-619, ISSN 0748-4658

Passrucker, H.; Engber, M.; Kablitz, S. & Hennecke, D. K. (2003). Effect of Forward Sweep in a Transonic Compressor Rotor, *Proc Instn Mech Engrs, Part A: Journal of Power and Energy*, Vol. 217, No. 4, (2003), pp. 357-365, ISSN 0957-6509

Prasad, A. (2003). Evolution of Upstream Propagating Shock Waves From a Transonic Compressor Rotor, *ASME Journal of Turbomachinery*, Vol. 125, No. 1, (January 2003), pp. 133-140, ISSN 0889-504X

Puterbaugh, S. L. & Brendel, M. (1997). Tip Clearance Flow-Shock Interaction in a Transonic Compressor Rotor, *Journal of Propulsion and Power*, Vol. 13, No. 1, (Jan-Feb 1997), pp. 24-30, ISSN 0748-4658

Puterbaugh, S. L.; Copenhaver, W. W.; Hah, C. & Wennerstrom, A. J. (1997). A Three-Dimensional Shock Loss Model Applied to an Aft-Swept, Transonic Compressor Rotor, *ASME Journal of Turbomachinery*, Vol. 119, No. 3, (July 1997), pp. 452-459, ISSN 0889-504X

Rabe, D. C. & Hah, C. (2002). Application of Casing Circumferential Grooves for Improved Stall Margin in a Transonic Axial Compressor, *Proceedings of ASME Turbo Expo 2002*, GT-2002-30641

Reid, L. & Moore, R. D. (1978). Design and Overall Performance of Four Highly Loaded, High-Speed Inlet Stages for an Advanced High-Pressure-Ratio Core Compressor, NASA TP 1337

Schuler, B. J.; Kerrebrock, J. L. & Merchant, A. (2005). Experimental Investigation of a Transonic Aspirated Compressor, *ASME Journal of Turbomachinery*, Vol. 127, No. 2, (April 2005), pp. 340-348, ISSN 0889-504X

Schuler, B. J.; Kerrebrock, J. L. & Merchant, A. (2002). Experimental Investigation of an Aspirated Fan Stage, *Proceedings of ASME Turbo Expo 2002*, GT-2002-30370

Strazisar, A. J. (1985). Investigation of Flow Phenomena in a Transonic Fan Rotor Using Laser Anemometry, *ASME Journal of Engineering for Gas Turbines and Power*, Vol.107, No.2, (April 1985), pp. 427-435, ISSN 0742-4795

Strazisar, A. J.; Wood, J. R.; Hathaway, M. D. & Suder, K. L. (1989). Laser Anemometer Measurements in a Transonic Axial-Flow Fan Rotor, NASA TP 2879

Strazisar, A. J.; Bright, M. M.; Thorp, S.; Culley, D. E. & Suder, K. L. (2004). Compressor Stall Control Through Endwall Recirculation, *Proceedings of ASME Turbo Expo 2004*, GT2004-54295

Suder, K. L. (1998). Blockage Development in a Transonic, Axial Compressor Rotor, *ASME Journal of Turbomachinery*, Vol. 120, No. 3, (July 1998), pp. 465-476, ISSN 0889-504X

Suder, K. L. & Celestina, M. L. (1996). Experimental and Computational Investigation of the Tip Clearance Flow in a Transonic Axial Compressor Rotor, *ASME Journal of Turbomachinery*, Vol. 118, No. 2, (April 1996), pp. 218-229, ISSN 0889-504X

Suder, K. L.; Chima, R. V.; Strazisar, A. J. & Roberts, W. B. (1995). The Effect of Adding Roughness and Thickness to a Transonic Axial Compressor Rotor, *ASME Journal of Turbomachinery*, Vol. 117, No. 4, (October 1995), pp. 491-505, ISSN 0889-504X

Suder, K. L.; Hathaway, M. D.; Thorp, S. A.; Strazisar, A. J. & Bright, M. B. (2001). Compressor Stability Enhancement Using Discrete Tip Injection, *ASME Journal of Turbomachinery*, Vol. 123, No. 1, (January 2001), pp. 14-23, ISSN 0889-504X

Sun, Y.; Ren, Y.-X.; Fu, S.; Wadia, A. R. & Wisler, D. (2007). Numerical Study of the Loss in a Compressor Stage, *Proceedings of the International Gas Turbine Congress 2007 Tokyo*, IGTC2007 Tokyo TS-041

Wadia, A. R. & Copenhaver, W. W. (1996). An Investigation of the Effect of Cascade Area Ratios on Transonic Compressor Performance, *ASME Journal of Turbomachinery*, Vol. 118, No. 4, (October 1996), pp. 760-770, ISSN 0889-504X

Wadia, A. R. & Law, C. H. (1993). Low Aspect Ratio Transonic Rotors: Part 2-Influence of Location of Maximum Thickness on Transonic Compressor Performance, *ASME Journal of Turbomachinery*, Vol. 115, No. 2, (April 1993), pp. 226-239, ISSN 0889-504X

Wadia, A. R.; Szucs, P. N. & Crall, D. W. (1998). Inner Workings of Aerodynamic Sweep, *ASME Journal of Turbomachinery*, Vol. 120, No. 4, (October 1998), pp. 671-682, ISSN 0889-504X

Weber, A.; Schreiber, H.-A.; Fuchs, R. & Steinert, W. (2002). 3-D Transonic Flow in a Compressor Cascade With Shock-Induced Corner Stall, *ASME Journal of Turbomachinery*, Vol. 124, No. 3, (July 2002), pp. 358-366, ISSN 0889-504X

Wennerstrom, A. J. & Puterbaugh, S. L. (1984). A Three-Dimensional Model for the Prediction of Shock Losses in Compressor Blade Rows, *ASME Journal of Engineering for Gas Turbines and Power*, Vol. 106, No. 2, (April 1984), pp. 295-299, ISSN 0742-4795

Weyer, H. B. & Dunker, R. (1978). Laser Anemometry Study of the Flow Field in a Transonic Compressor Rotor, *Proceedings of AIAA Aerospace Sciences Meeting 1978*, AIAA Paper 78-1

Wilke, I. & Kau, H.-P. (2004). A Numerical Investigation of the Flow Mechanisms in a High pressure Compressor Front Stage With Axial Slots, *ASME Journal of Turbomachinery*, Vol. 126, No. 3, (July 2004), pp. 339-349, ISSN 0889-504X

Yamada, K.; Furukawa, M.; Nakano, T.; Inoue, M. & Funazaki, K. (2004). Unsteady Three-Dimensional Flow Phenomena Due to Breakdown of Tip Leakage Vortex in a Transonic Axial Compressor Rotor, *Proceedings of ASME Turbo Expo 2004*, GT2004-53745

Yi, W.; Huang, H. & Han, W. (2006). Design Optimization of Transonic Compressor Rotor Using CFD and Genetic Algorithm, *Proceedings of ASME Turbo Expo 2006*, GT2006-90155

Part 2

Gas Turbine Systems

4

Biofuel and Gas Turbine Engines

Marco Antônio Rosa do Nascimento and Eraldo Cruz dos Santos
Federal University of Itajubá – UNIFEI
Brazil

1. Introduction

Currently, the interest in using vegetable oils and their derivatives as fuel in primary drives for the generation of electricity has increased due to rising oil prices and concerns over the environmental impacts caused by fossil fuel use. For viability of using biodiesel as a substitute for fossil fuels for power generation, should be considered the emissions of greenhouse gases, i.e., pollutants such as nitrogen oxides (NO_X), sulfur oxides (SO_X), carbon monoxide (CO) and particulates into the atmosphere during the lifetime of the power plant.

According HABIB (2010) the effect of using petroleum-derived fuel in aviation on the environment is significant. Given the intensity of air traffic and civil and military operations, making the development of alternative fuels for the aviation sector is justified, necessary and critical.

Another concern that must be considered is the quality of biofuel to be stored over time, this being an obstacle to be overcome in order to maintain fuel quality and operational reliability of gas turbine installations operating with biofuels.

Biofuels also have the advantage of being renewable and cleaner, this is due in large part because they do not contain sulfur in its composition. The use of distributed generation renewable fuels can be advantageous in isolated regions, far from major urban centers, to generate electricity using the resources available on site.

Among other engines, gas turbines represent one of the technologies of distributed generation, which is characterized by the supply of electricity and heat simultaneously. In principle these machines should operate without major problems by using biofuels, because of similarities with the characteristics of the fuels conventionally used. However, there are few references on the performance of gas turbines operating on biofuels and this is the motivation of this study.

Microturbines are small gas-turbo generators designed to operate in the power range from 10 to 350 kW. Although its operation will also be based on the Brayton cycle, they present their own characteristics that differentiate them from large turbines.

Most gas turbine available today, originated in the military and aerospace industry. Many projects were aimed at applications in the automotive sector in the period between 1950 and 1970. The first gas turbine generation was developed from turbo aircraft, buses and other commercial means of transport (SCOTT, 2000). Interest in stationary generation market has expanded in the years 1980 and 1990, and its use in distributed generation has been accelerated (LISS, 1999).

It is hoped that in future gas turbines of small power is an alternative for power generation for residential and commercial segment, since the operational reliability is one of the main needs in these sectors (WILLIS and SCOTT, 2000). These turbines have various applications such as power generation in the place of consumption (on-site), the uninterrupted supply of electricity, to cover peak loads, cogeneration and mechanical drive, which characterizes the distributed generation (BIASI, 1998).

Gas turbines may use different types of fuel such as diesel, kerosene, ethanol, natural gas and gas obtained from biomass gasification, etc. The shift to gas from biomass has been considered promising, but some changes must be made in supply and combustion systems turbine, aiming to modify the injection and control systems and the volume of the combustion chamber.

The scope of this chapter includes a brief description of the systems of gas turbines, reports the experiences of biofuel use in gas turbines made until today, with emphasis on the experience developed in the Laboratory of Gas Turbines and Gasification the Institute of Mechanical Engineering, Federal University of Itajubá - IEM/UNIFEI aspects of thermal performance and emissions of gases from gas turbines of small power.

2. Biofuels

Biofuels are fuels of biological origin, i.e., not fossil. They are produced from plants such as corn, soy, sugar cane, castor beans, sugar beet, palm oil, canola, babassu oil, hemp, among others. Organic waste can also be used for the production of biofuel. The main biofuels are ethanol (produced from sugar cane and corn), biogas (biomass), bioethanol, biodiesel (from palm oil or soy), among others.

Biofuels can be used on vehicles (cars, trucks, tractors, etc.), turbines, boilers, etc.., in whole or blended with fossil fuels. In Brazil, for example, soy biodiesel is blended with fossil diesel. Is also added to gasoline the ethanol produced from sugar cane.

The advantage of using biofuels is the significant reduction of greenhouse gas emissions. It is also advantageous because it is a renewable source of energy instead of fossil fuels (diesel, gasoline, kerosene, coal).

This section will describe some characteristics and requirements of biofuels that have potential for use in gas turbines.

2.1 Gas turbines operating on liquid fuels

Biofuels have the greatest potential for use in gas turbines are biodiesel and ethanol, due to factors such as availability physical-chemical characteristics similar to fossil fuels such as diesel or jet fuel. Table 1 presents a summary of requirement of liquid fuel as defined by the manufacturers of gas turbines for efficient operations (BOYCE, 2006).

Moisture and sediment	1.0 % (v%) maximum
Viscosity	20 cS at injector
Dew-Point	20 °C at ambient temperature
Carbon Residue	1.0 % (p.) maximum
Hydrogen	11% (p.) maximum
Sulfur	1% (p.) maximum

Table 1. Requirements liquid fuel for gas turbines.

The growing interest in biofuels along with increasing market demand for generators supplied by renewable fuels has led manufacturers to modify the designs of gas turbines and micro-turbines, in order that they can operate on biofuels.

For biodiesel, the supply system is being modified to fit this new biofuel due to some reasons such as higher viscosity, content of acylglycerols and the effects of corrosion. New corrosion-resistant materials, systems control the flow of fuel and improved geometry optimized for the guns are some of the challenges of these new projects.

In scientific literature there is little information about testing of gas turbines for small power, operating on biofuels. To study the impact of biofuel use in the operation and maintenance of gas turbine, one must take the following measures:

- Define the physical and chemical characteristics of both diesel and biofuel used in the tests. Some important characteristics are: density, distillation, viscosity, ash content, phosphorus, iodine and sulfur, water content, cetane number, oxidation stability, flashpoint, freezing point, dew point, volumetric composition of methyl, ethyl and lipids, glycerol, lower calorific value, etc. These values should be compared with the requirements of the standards on diesel and biodiesel to demonstrate that they can be used in the study.

- Once further tests it is possible to define what characteristics of biodiesel are relevant to the determination of changes in engine behavior, considering the performance parameters and emissions. The higher viscosity of biodiesel can lead to difficulties in its injection into the combustion chamber. It is possible reduce the viscosity of the mixture increases its temperature, or by adding alcohol. The lower the flash point of biodiesel could also cause problems in combustion.

- It is possible find accumulation of carbonized material in the inner parts of the gas turbine, after the tests with biodiesel. The biofuel can produce corrosion in fuel supply system.

- It is also recommended to install a filter at least 50 µm at the fuel supply in the gas turbine when using biodiesel.

- It is not advisable to use biofuels in gas turbines without performing a preliminary economic analysis.

Some alternative liquid fuels such as vegetable oil, biodiesel or pyrolysis oil, ethanol and methanol are being tested in gas turbines (GÖKALP, 2004).

As biodiesel has similar properties to diesel, it can be used directly in a gas turbine, blended with diesel in various proportions (usually uses 5 to 30% biodiesel in the blend with diesel). The properties of biodiesel are slightly different to those of diesel in terms of energy content or physical properties.

The Lower Heating Value (LHV) of liquid biofuels such as pure biodiesel (B100), B5 B30 and vegetable oils are between 37,500 and 44,500 kJ/kg, which is close to regular diesel (GÖKALP, 2004).

The viscosities of ethanol, biodiesel and its blends with diesel are lower than the residual oil from the kitchen, making it easier to spray. Vegetable oils and oils derived from pyrolysis have a very high viscosity, which causes problems in its mist inside the combustion chamber of gas turbines. However, these fuels can be heated to reduce its viscosity before being injected into the combustor.

Fuel oil resulting from pyrolysis of wood, vegetable oils and methyl esters has a carbon/hydrogen rate (C/H) higher than that of conventional diesel. As consequence, there may be an accumulation of soot inside the combustion chamber or turbine blades (GÖKALP, 2004). Another factor that changes as a result of this feature is the transfer of heat by radiation from the flame to the flame tube.

2.2 Gas turbines operating on gaseous fuels

Thermal power plants with gas turbines operating with gas from biomass need to present efficient, simple technology, low cost and operational reliability, in order that these plants could become economically competitive with traditional systems of power generation, such as stationary alternative engines.

The potential of gas turbines for this application is great, although the gas must be subjected to cleaning to remove solid impurities and/or gas that can damage some components of some systems of gas turbine.

Gaseous fuels can be obtained by gasification of biomass, which in addition to the gas generates a set of substances, such as tar which is a compound in gaseous form in the fuel gas, which has an appreciable calorific value, although it shows a tar obstacle to the use of gaseous fuels in internal combustion engines, due to its high corrosive power of components and reduced engine efficiency.

In the case of gas turbines, the tar can be a problem only happens when its condensation. Basically there are two strategies to address the problem of tar, remove it from the fuel gas or burn it in the combustion chamber. In the first case, nickel-based catalysts have shown very promising results. In the second case, the strategy is to keep the fuel temperature above the dew point of tar in the gas supply pipes, and perform your burning at high temperature in the combustion chamber (SCHMITZ, 2000).

Currently, gas turbines are designed for a specific fuel (natural gas or fuel oil). Recent progress has been achieved in the methodologies and tools for the design of combustors for gas turbine. It is possible to perform a clean combustion of fossil fuels by employing low-carbon technologies based on premixed combustion.

There are ongoing projects that aim to harness these advances for applications geared to a wider range of fuels with commercial potential, including those with low calorific value, obtained from biomass gasification. Some procedures should be established for selecting appropriate fuels to be used taking into account the performance of combustion and emissions of soot and NO_X. Furthermore, it should be considered the adaptability of existing burners to use alternative fuels selected (GÖKALP, 2004).

Some gaseous alternative fuels also have potential for use in gas turbines, for example, the synthesis gas from gasification of biomass, the biomass pyrolysis gas, gas from digesters (biogas) and residual gas from industrial processes, which are rich in hydrogen.

Industrial gases such as methane reformed with steam, refinery gas, residual gas from the Fischer-Tropsch gasification gas with oxygen gas and slow pyrolysis of wood, have a LHV comparable to natural gas. This is due to the high hydrogen content of fuel gases, which lies between 19 and 45% of the volume. Rather, the LHV of gas gasification with air and biogas are very low because they are produced at atmospheric pressure, so they must be compressed before being used in gas turbines.

Except reformed methane with steam, all other gaseous fuels mentioned above have a C/H greater than that of natural gas (GÖKALP, 2004).

According BOYCE (2006) in Table 2 presents an overview of the requirements for gaseous fuels that can be used in gas turbines.

Calorific	11,184 – 41,000 kJ/m³
Solid Contaminants	< 30 ppm
Flammability Limits	2.2 : 1
Content of sulfur, sodium, potassium and lithium	< 5 ppm (In the form of meta alkaline sulfate)
H₂O (p.)	< 25%

Table 2. Requirements gaseous fuel for gas turbines.

According KEHLHOFER (2009) the type and composition of the fuel has a direct influence on efficiency and emission of gases from a gas turbine. The LHV of the fuel is important because it defines the mass flow of fuel and consequently its specific consumption. The fuel composition is also important as it influences the performance of the cycle because it determines the enthalpy of gas entering the turbine, and the available enthalpy drop off the engine and the amount of steam generated in the recovery boiler.

2.3 Clean fuel gas

Gaseous fuels obtained through the processes of gasification and pyrolysis produce fuels with low-and medium calorific value. According to Table 1, in some gas turbines require a minimum value for the calorific value of gas, which can be difficult to achieve it. Manufacturers set very strict parameters regarding the quality of the combustible gases to avoid possible damage to the hot parts of gas turbine. Alkaline components present in the fuel, especially chlorides cause corrosion at high temperatures. The presence of tar should also be considered if the operations of fuel valves occur at temperatures below the dew point of the tar. Almost all biogenic fuels contain considerable amounts of halogens (Cl, F, Br), alkali (Na, K), alkali oxides and other metals (Zn, Cu, Ca). The sulfur content is usually low. Unless most of these items may be retained in the gasifier or pyrolysis reactor, the requirements of the fuel gas cannot be met when employing biomass. The fuel gas contains impurities mentioned in the gaseous or solid, so it is inevitable that they would perform a deep cleaning of the gas. The chloride concentration should be considerably reduced. The slow pyrolysis produces gases, which are characterized by average values of calorific value, situated in the range between 7,000 and 13,000 kJ/kg, and low levels of chlorides and alkalis.

Should be applied under heating rates and moderate temperatures to produce pyrolysis gas with low alkali chloride. High rates of heating fuel break biogenic structure, which favors the conversion of solid pyrolysis gas, however, compromises the retention of impurities in the vegetable coal. High levels of retention in coal are only achieved with low rates of heating in combination with moderate temperatures.

It is estimated that for slow pyrolysis with a maximum temperature of 350 °C are retained approximately 86% of chloride, and the calorific value of gas reaches 10.9 MJ/kg, which can be considered average. These conditions result in a lower ratio chloride/energy in the gas, although the concentrations required by gas turbines today are even smaller, and therefore, solid and gaseous impurities must be removed (SCHMITZ, 2000).

3. Experiments with biofuels in gas turbine engines

This item will be described some experiences with the utilization of biofuels in gas turbines. Testing gas micro-turbines operating with biodiesel were performed by LOPP et. al. (1995); MIMURA (2003); BIST (2004); SCHMELLEKAMP and DIELMANN (2004); WENDIG (2004) and CORRÊA (2006). LOPP (1995) presented the results of the thermal performance of a gas turbine using a blend of jet fuel (Jet fuel - JF) and soy biodiesel (B). Three different fuels: JF, B10/JF90 and B20/JF80. The turbine efficiency reached its nominal value with B20, with a slightly better performance than with B10. There was an increase in fuel consumption proportional to the addition of biodiesel in the blend. The blends with biodiesel/diesel fuel were compatible enough to allow additional testing and show its potential as an alternative fuel. CO_2 emissions were reduced after the engine was fueled with B20. The two blends B10 and B20 did not show a noticeable increase in emissions of particulate matter compared to the JF. Subsequent inspections in the combustor and turbine blades showed no deposition or degradation of components.

MIMURA (2003) conducted a study of performance and emissions from a micro-turbine supplied with biodiesel from waste food oil regenerated. Operating in cogeneration mode the system had a thermal efficiency of 64%. It was observed emissions of 6 ppm of CO, 23 ppm of NO_X and 1.0 ppm of SO_X.

BIST (2004) performed a feasibility study on the use of methyl esters (biodiesel) derived from soy oil as additives blended with fuel gas turbine aircraft. Several blends were tested to identify which would meet the specifications for this type of engine without the need to change the initial design. Was not noticed a significant increase in fuel consumption for blends of B2, B5 and B10. In cases of B20 and B30 blends the increase in consumption was evident, being 7 and 10%, respectively. It was shown that a decrease in the efficiency of combustion in gas turbine as the percentage of biodiesel in the blend increase. In terms of emissions, an increase of CO content in the gases, due to the increase in the percentage of biodiesel in the blend, was noticed what indicates a reduction in combustion efficiency. This behavior is contrary to that seen in piston engine in which the CO decrease with increasing content of biodiesel blends.

There was also an increase in emissions of NO by increasing the concentration of biodiesel, therefore, NO_2 emissions did not change significantly. The high viscosity of biodiesel is a limiting factor for not testing with larger percentages than B30.

The biodiesel from soybeans contain glycerin. The amount of glycerin in the mixture should be kept as low as possible to avoid problems in combustion process. These measurements indicated that the B30 blend showed a content of glycerin which can be considered negligible. In none of the blends tested has been observed increases in pressure drop through analysis in the fuel filter and therefore none of them produced sediments that could cause blockage in the supply system of the turbine engine (BIST, 2004).

SCHMELLEKAMP and DIELMANN (2004) present the results of using vegetable oil from rape seed in a 30 kW micro-turbine, in blends of 10, 20 and 30%. Fuel consumption increased with increasing biodiesel blend. Using B30 obtained a 12% higher consumption in the range of operation. By using B10 it was verified that CO were lower than in the case of use of fossil diesel. However, when were employed mixtures of B20 and B30 it was observed a higher level of CO emissions. The viscosity of vegetable oil is much greater than that of biodiesel, so it is necessary to make a preheating the blend prior to injection.

The results of operating a 75 kW micro-turbine, burning biodiesel from rapeseed, sunflower, animal fats, were presented by WENDIG (2004). The operation with these three kinds of methyl esters showed a significant increase in emissions of CO and CO_2 at full load. All fuels examined showed a reduction in NO_X emissions in the range of 55%. Were problems related to the corrosive characteristics of biodiesel.

More recently, tests of performance and emissions in a 30 kW micro-turbine, using biodiesel from castor beans, were published by CORRÊA (2006). During the tests it was necessary to preheat the blends to 40 °C to achieve the viscosity values required by the manufacturer. The specific fuel consumption increased nearly 21% when using B100. Throughout the power range studied, it was observed a reduction in emissions of CO and NO_X in the exhaust gases.

WENDIG (2004) by using biogas found a decrease in CO with increasing load, reaching 100 ppm at full load. Since the emission of CO_2 was about 45 ppm, constant throughout the range of operation, and SO_2 emissions decreased. At full load the SO_2 emission was practically zero. The micro-turbine showed the lowest emission of NO_X at full load (20 ppm).

The existing reports on the use of biodiesel in micro-turbines describe increases in fuel flow rate when increasing the proportion of biodiesel in the blend; this is explained in part by lower LHV of biodiesel compared to diesel.

Problems due to viscosity, corrosion and accumulation of foreign material in the turbine blades may be in extended operations. These problems can be mitigated through a rigorous verification of the characteristics of biodiesel, demanding that it complies with the standards and manufacturers recommendations. In Brazil, ABNT published standards NBR 15341, NBR 15342, NBR 15343 and NBR 15344, to specify the properties of biodiesel (ABNT, 2006).

Manufacturers of gas turbines and micro-turbines are currently developing models that can use biofuels such as biodiesel. The changes that are taking place mainly consist in the use of materials resistant to corrosion in fuel and injection systems, and develop systems adjusted to work with injection of fuel physical-chemical characteristics different from diesel.

4. Cycles with potential for use of biofuels

A gas turbine is a set of three components: the compressor, combustion chamber and the turbine itself. This configuration forms a gas thermodynamic cycle accordance with the model ideal Brayton cycle is called. This set can operate in an open cycle, with air as the working fluid, which is admitted to the atmospheric pressure, passes through the turbine and is discharged back into the atmosphere without returning to the admission.

Thus, despite being an open cycle, some energy from the combustion is rejected in the form of heat contained in hot exhaust gases. The heat rejection is a physical limit of the gas turbine, intrinsic to the operation of thermodynamic cycles, even in ideal cases.

The loss of condition ideal cycle in a gas turbine can be quantified by the ratio involving the calorific value of fuel, discounting the power to drive the compressor and power net. Thus, decreasing the losses as it reduces the exhaust temperature, and raises the temperature of turbine inlet. The resistance to high temperature components of gas turbine is a very critical point in building technology such equipment (COHEN, H. et. al., 1996).

Turbines designed to operate in simple cycle, as shown in Figure 1, in view of the thermal efficiency of the cycle, have gas outlet temperature reduced to maximum and have optimized compression ratio. The compression ratio is the ratio between the pressure of air entering and exiting the compressor. For example, if air enters at 1.0 atm, and leaves the compressor at 15.0 atm, the compression ratio is 15:1.

Apart from variation of the simple cycle obtained by the addition of these other components, considerations must be given to two system distinguished by the use of open (Figure1) and close cycles (Figure 2).

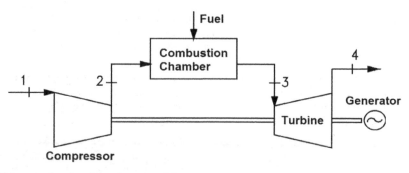

Fig. 1. Scheme of a simple cycle gas turbine system.

In the process of closed cycle operation is the same as the open cycle, the difference is that the working fluid remains within the system and the fuel is burned outside the system.

The biggest advantage of the closed cycle is the possibility of using high pressure throughout the circuit, which results in reducing the size of the turbomachinery, depending on power output, and allows the variation of power output by varying the pressure level of the circuit.

The significant feature is that the hot gases produced in the boiler furnace or reactor core never reach the turbine; they are merely used indirectly to produce an intermediate fluid, namely steam.

In order to produce an expansion through a turbine a pressure ratio must be provided and the first necessary step in the cycle of gas turbine plant must therefore be compression of the working fluid.

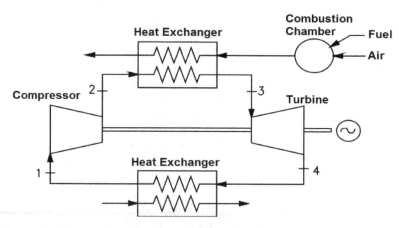

Fig. 2. Scheme of a close cycle gas turbine with a heat exchanger.

If after compression the working fluid was to be expanded directly in the turbine, and there were no losses in either component, the power developed by the turbine would just equal that absorbed by the compressor. Thus if the two were coupled together the combination would do no more than turn itself round. But the power developed by the turbine can be increased by the addition of energy to raise the temperature of the working fluid prior to expansion. When the working fluid is air a very suitable means of doing this is by combustion of fuel in the air which has been compressed.

One way to increase the thermal efficiency of a closed cycle is the addition of a heat exchanger; however there are limits to the introduction of this equipment because it can cause loss of pressure in the circuit.

A cycle gas turbine with a heat exchanger can also be called a regenerative cycle, as shown in Figure 3, wherein the heat rejected in the exhaust gases of the gas turbine passes through the heat exchanger and heats the air from the compressor before entering the chamber combustion. The pre-heated air reduces the fuel consumption injected into the combustion chamber, increasing the thermal efficiency of the cycle.

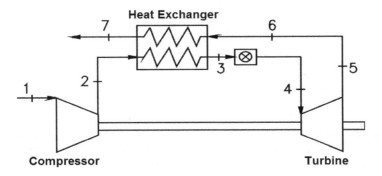

Fig. 3. Scheme of a regenerative cycle gas turbine.

Specific gas turbines to operate in combined cycle, as shown in Figure 4, are developed in order to maximize the thermal efficiency of the cycle as a throughout. Therefore, reducing the temperature of the exhaust gas is not necessarily the most critical point in terms of efficiency, since the gas turbine exit are still used to generate power in other equipment.

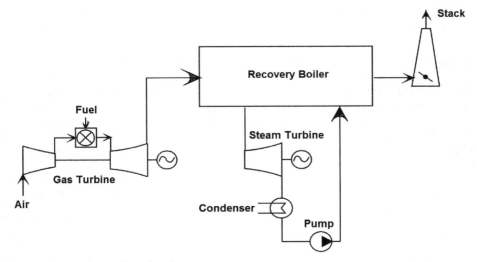

Fig. 4. Scheme of a combined cycle.

Another configuration of installation of gas turbines is the cycle where there is the presence of an intercooler before or between compressors, i.e., compressors among the low and high pressure. In this configuration the air that enters into the first compressor is compressed to a pressure intermediate between the maximum of the cycle and the ambient pressure, as shown in Figure 5.

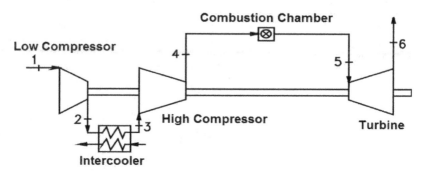

Fig. 5. Scheme of a cycle with intercooler.

In the case of plants with two compressors the air leaving the first compressor (low pressure) enters the intercooler where the heat is removed without major pressure drop. Being that, in practice, the air is returned to the circuit should have a temperature slightly higher than the entrance to the first compressor.

In this type of gas turbine is possible to operate with temperatures above 1450 °C due to the reduction of air temperature of cooling the turbine blades, increasing the thermal efficiency of the cycle, reaching 50%.

The combined cycle consists of one or more gas turbine, whose exhaust gases are injected into a recovery boiler that provides steam to a turbine.

In an open cycle, the thermal efficiency is low, around 30 to 35%. In a combined cycle efficiency can reach 60% (the highest efficiency of all types of driver). In combined cycle occurs the combination of gas turbines with steam turbines.

4.1 Selection cycle gas turbine during operation with gas biofuel

In simple cycle the thermal efficiencies is increased with the increase of pressure ratio and turbine inlet temperature. Therefore, high pressure ratios require high pressure to clean the gas, which results in high costs of equipment, and high compression energy consumption.

In a cycle of regeneration is evident with a peak thermal efficiency at a pressure ratio between 4 and 8. In relation to biomass gasification, the cycle with regenerator is interesting because it is possible to use moderate pressure gasification resulting in lower equipment costs and acceptable thermal efficiencies compared to the simple cycle.

Preheating inlet air aerator does not increase much thermal efficiency, since the output of the reactor; higher temperatures are derived fuel gas. However, this represents a contradiction, since the positive effects of preheating are reversed, it is necessary to perform the cooling of the gas for cleaning.

With gas pyrolysis cycle gas turbine with a regenerator would lower overall efficiency, because the residual heat from the pyrolysis process cannot be used to increase the power of the turbine.

The combined cycle integrated with slow pyrolysis uses a Rankine cycle supplied with charcoal and a cycle gas turbine powered by pyrolysis gas, which results in a high cycle efficiency, because a considerable amount of waste heat can be used to produce Rankine cycle power (SCHMITZ, 2000).

4.2 Adjustments to gas turbine

Due to impurities in the gas or fuel, for example, the synthesis or biofuel, it is necessary redesign the gas turbine combustor. For each type of fuel, run a kind of optimization, with reference to a low value of the LHV of fuel.

To compensate for the lower value of LHV for the fuel gases, the fuel injection system must provide a fuel rate much higher than when the combustor operates fuel with high calorific value. Due to the high rate of mass flow of gas with low LHV, the passage of fuel has a much larger cross section than the section corresponding to natural gas. The fuel pipes and control valves and stop valve have larger diameters and shall be designed to include an additional fuel blend, which consists of the final mixture of the recovered gas with natural gas and steam. The pressure drop and the size of the spiral of air entering the flame tube were adjusted to optimize the combustion process. The system must have high safety standards, so the flanges and gaskets of the combustor and its connections must be good soldiers. The system for low fuel LHV must include:

- Fuel line with a low LHV;
- Natural gas line;
- Steam line to reduce NO_X;
- Line blending of fuel with low LHV;
- Line of nitrogen to purge;
- Lines pilot;
- Compressor;
- Combustion Chamber.

The loading of the gas turbine to the rated load is accomplished through the use of the fuel reserve for security reasons. The procedure for replacing the fuel reserve for the main runs automatically. The characteristics of blends are monitored and analyzed online.

5. Case study with biodiesel and ethanol in gas turbine

For perform the tests in the gas turbine was built a test bench in the laboratory of gas turbines and gasification of the Institute of Mechanical Engineering, Federal University of Itajubá - IEM/UNIFEI. This bench is made of a micro-turbine Capstone C30 model cycle with regenerative power of 30 kW, configured to operate with liquid fuel.

This gas turbine is used mainly for primer power generation or emergency and can work in a variety of liquid fuels. This turbine uses a recovery cycle to improve its efficiency during operation, due to a relatively low pressure what facilitates the use of a single shaft radial compression and expansion (BOLSZO, 2009).

The experimental tests were carried out by using a 30 kW regenerative cycle diesel single shaft gas turbine engine with annular combustion chamber and radial turbomachineries, whose characteristics at ISO conditions are given in Table 3 (CAPSTONE, 2001).

To perform the experiments with different fuel blends and it was implemented a system to preheat the fuel supply aiming to control the viscosity of the fuel used.

For tracking and measuring the parameters of the test was used a supervisory software in test bench (given by the manufacturer of the turbine) and a data acquisition and post processing of data obtained during the tests.

Fuel Pressure	350 kPa
Power Output	29 kW NET (± 1)
Thermal Efficiency	26% (± 2)
Fuel HHV	45,144 kJ/kg
Fuel Flow	12 l/h
Heat Rate (LHV)	14,000 kJ/kWh
Exhaust Temperature	260 °C
Inlet Air Flow	16 Nm³/min
Rotational Speed	96000 rpm
Pressure Ratio	4

Table 3. Engine Performance data at ISO Condition.

5.1 Tests on gas turbine using biodiesel fuels

To perform the experimental tests were used blends biodiesel/diesel, including the total replacement of diesel with biodiesel in the gas turbine. The blends considered in the experiment were: B10, B20, B30, B50 and B100. Due to low solubility in diesel fuel at low temperature tests with ethanol were performed without pre-mix, and also without the use of additives, which enhance the cost of fuel.

Following the methodology of the measures to be adopted to test gas turbine (item 2), Table 4 shows the physical-chemical properties of biodiesel for testing thermal performance and emissions:

Properties	Soy Biodiesel	Palm Oil Biodiesel	Diesel	Micro Turbine Manufacturer Fuel Limits	ASTM D6751
Cetane Number	58	60	45.8	-	> 47
Sulfur (% mass)	0	0	0.20	0.05 <	< 0.05
Kinematic Viscosity @ 40 °C (mm²/s)	2.19	2.26	1.54	1.9 – 4.1	1.9 – 6
Density @ 25 °C (g/cm³)	0.888	0.854	0.838	0.75 – 0.95	-
Flash Point (°C)	136	138	60	38 - 66	> 130
Water (% Volume)	0.05	0.05	0.05	0.05	0.05

Table 4. Biodiesel and diesel physic-chemical characteristics.

Table 4 shows the comparison amongst the characteristics of commonly used types of biodiesel and diesel pure, with the requirements of the fuel made by the manufacturer of gas turbine tested and specifications for biodiesel fuel blend of standard ASTM D6751.

Once obtained the blends, there was the experimental determination of the calorific value, kinematic viscosity and density of different blends (B10, B20, B30, B50, B100) and ethanol, according to the standards ISO 1928-1976 and ASTM D1989-91, respectively. In the case of viscosity were made ten measurements to reduce the standard deviation, and the calorific value were performed five measurements for each blend. Table 5 shows the LHV of the fuels used in the experiment.

Fuel	Pure (kJ/kg)	B10 (kJ/kg)	B20 (kJ/kg)	B30 (kJ/kg)	B50 (kJ/kg)
Biodiesel (Palm Oil)	37,230.37	41,204.29	40,691.27	39,768.17	38,623.27
Biodiesel (Soy)	28,298.16	41,425.27	40,748.82	39,864.36	39,061.45
Diesel	42,179.27	-	-	-	-
Alcohol	23,985.00	-	-	-	-

Table 5. Lower Heating Value of fuels used.

The use of different fuels implies the need to make adjustments of mass flow rate of them, according to its LHV and your density, because without these adjustments, once established a load, the supply system would feed a quantity of fuel depending on the characteristics of standard fuel (diesel). If the LHV of the new fuel is lower than the standard, the gas turbine power could not reach the demanded.

Initially, the engine was operating with conventional diesel fuel for a period of 20 minutes to reach a steady state condition for a load of 10 kW. After 20 minutes, the mass flow rates were changed to the corresponding values of blends diesel/biodiesel. At this stage, it begins to replace the fuel, in order of increasing content of biodiesel (B10, B20, B30, B50 and B100), closing the inlet valve of diesel and opening the valve of the mixture. In order to ensure that all existing diesel power on the internal circuitry of the engine would be consumed, the engine was left operation for 10 minutes with the same load operation (10 kW).

In order to check if the fuels were able to feed the engine, without experiencing any problems, regarding the fuel injection system, the kinematic viscosity of each fuel was measured. The composition of gas emissions and thermal parameters were also measured in total and average load for each fuel. This whole procedure was performed for the engine operating with loads of 5, 10, 15, 20, 25 and 30 kW in a grid connection mode.

Afterwards it was held on measurement of emissions with gas analyzer, and increased the load of 5 kW, 10 minutes waiting again until it reaches steady state again.

When finished testing with a blend, the engine was left running, in order to accomplish the purging of fuel remaining. After executing the purge the supply system it was loaded with a new mixture. Once completed the tests with biodiesel and blends, was returning to operate the engine with diesel for ten minutes, and then it was disconnected and stopped.

5.2 Thermal performance

The results of performance testing of a 30 kW gas turbine engine supplied with biodiesel from palm oil, soy and ethanol are shown:

Figures 6, 7 and 8 show the relationship between specific fuel consumption and power. The graphs correspond to soy biodiesel, Figure 6, biodiesel from palm oil, Figure 7 and diesel and ethanol, Figure 8.

In the case of biodiesel from soy, Figure 6 observed if an increase in specific fuel consumption, when increasing the fraction of biodiesel in the blend in the range of 10 to 25 kW. The lowest value occurs when using pure diesel oil, the highest value occurs when using pure biodiesel and the difference between the curves of diesel and B100 remains approximately constant at 12.82%. This behavior if repeated when using biodiesel from palm oil as shown in Figure 7. When using ethanol, Figure 8 also shows an increase in specific fuel consumption with respect to diesel.

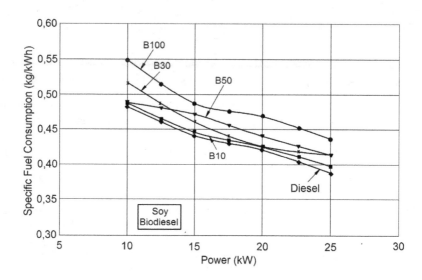

Fig. 6. Specific fuel consumption versus power of the gas turbine for different fuels: soy biodiesel.

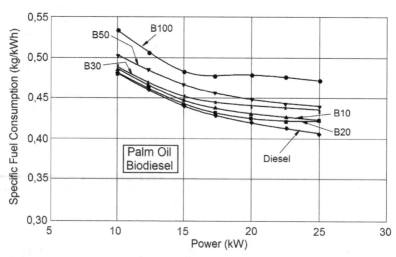

Fig. 7. Specific fuel consumption versus power of the gas turbine for different fuels: palm oil biodiesel.

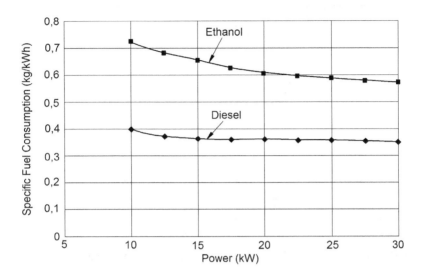

Fig. 8. Specific fuel consumption versus power of the gas turbine for different fuels: diesel and ethanol.

The specific consumption of biodiesel from soy and palm oil were approximately equal. Consumption already of ethanol was higher due to even lower calorific value of fuel compared with the others used fuel.

Figures 9, 10 and 11 show the behavior of power versus Heat Rate of gas turbine for the three different fuels tested.

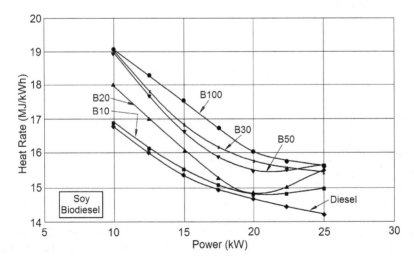

Fig. 9. Heat Rate versus power for different fuels: soy biodiesel.

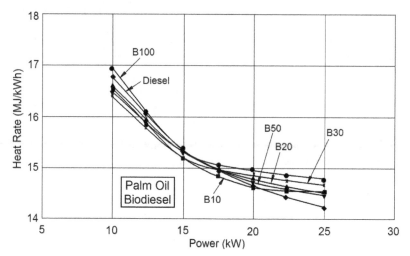

Fig. 10. Heat Rate versus power for different fuels: palm oil from biodiesel.

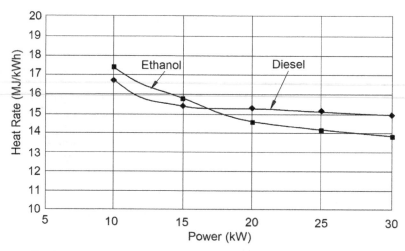

Fig. 11. Heat Rate versus power for different fuels: diesel and ethanol.

In all cases, it was observed a reduction in Heat Rate with the increase in power. Operation with biodiesel from palm oil presented a better performance than soy biodiesel (lower value of the Heat Rate). Similarly to the specific fuel consumption, the differences in the amount of Heat Rate between diesel and B100 remained approximately constant in the range 10 to 25 kW. The lower fuel consumption occurs to the rated power when operating with diesel. This is due to the higher calorific value of diesel.

When a lower LHV of fuel is used, a greater mass of fuel is needed to release in the combustion energy required for a specific power. The mass flow rate of fuel passing through turbine increases and the compressor operating point changes, making their efficiency and, consequently, the cycle efficiency decreases.

Finally, Figures 12, 13 and 14 displays the graphs of the thermal efficiency of the gas turbine for different fuel and power operation. The differences between the efficiency with diesel and biodiesel blends with soy biodiesel, Figure 7 may be due to differences in density, viscosity and LHV of fuel compared with conventional biodiesel. This can cause changes in the process of atomization of the fuel within the combustion chamber, reducing the thermal efficiency of the engine as previously mentioned. In the case of palm oil biodiesel was not observed differences, and the efficiency values remain equal to those obtained when the engine was tested with diesel, within the entire power range evaluated. Ethanol showed the lowest efficiency among the fuels tested

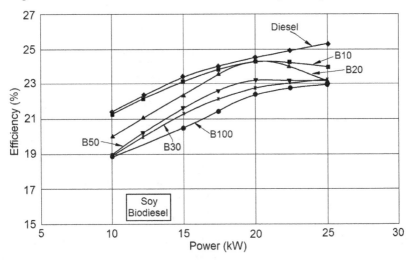

Fig. 12. Efficiency versus power turbine for different fuels: soy biodiesel.

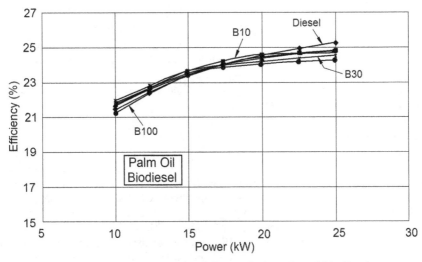

Fig. 13. Efficiency versus power turbine for different fuels: palm oil biodiesel.

As in tests performed by HABIB (2010), the test conducted in the laboratory of UNIFEI with B100 biodiesel resulted in high thermal efficiency compared to other blends, as shown by the graphs of Figures 12 and 13, such performance is attributed to equivalence ratio, which produced the best ratio of air to fuel during firing, resulting in more complete combustion due to the presence of extra oxygen in the biofuel, which resulted in the presence of 16 to 19% oxygen in the gases turbine exhaust.

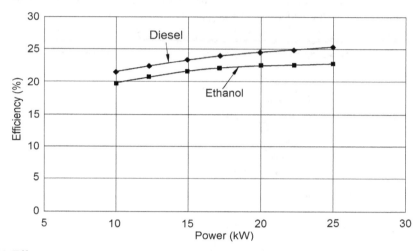

Fig. 14. Efficiency versus power turbine for different fuels: diesel and ethanol.

As mentioned by other authors, it can say that the type of biodiesel is an important factor in the analysis of projects for power generation, as small differences in density, viscosity and LHV cause changes in the parameters of thermal performance of turbines gas, and amongst them when compared with pure diesel.

For the case of palm oil biodiesel, although there is an increase in specific fuel consumption, there were no significant differences in the efficiency of the engine. With soy biodiesel there were differences in the efficiency of around 2% throughout the power range tested.

The results obtained with ethanol were very different for both types of biodiesel tested, due to its lower calorific value.

In practical terms, for distributed generation, ethanol must go through economic and technical assessments in order to detect.

5.3 Emission of pollutants

Publications on experiments with biofuels in gas turbines are not yet sufficient to make definitive conclusions in terms of emissions. However, it is observed that the CO decreases with increasing load. The opposite happens with the NO_X emissions. It is also provided a reduction in the emission of smoke, along with the increase in the emission of NO_X.

The emission results shown in the following are results of experiments with three different biofuels: Soy biodiesel, palm oil biodiesel and ethanol achieved in the laboratory of UNIFEI. The first two were blended with diesel in different proportions and each mixture was tested with a gas micro-turbine operating at full load and partial loads. The pure ethanol was used and compared with the performance of pure diesel, as reported in the results performance.

The analysis focuses mainly on changes in the levels of carbon monoxide (CO) and nitrogen oxides (NO$_X$), unburned hydrocarbons were not detected in the combustion products, during the tests. This is due to high temperatures and high excess air into the combustion chamber. Likewise there were no emissions of sulfur oxides (SO$_X$), since biofuels evaluated did not contain sulfur in your composition.

Figures 15, 16 and 17 shows the concentrations of CO in the exhaust gas from the micro-turbine gas to biodiesel blends: soy biodiesel, Figure 15, palm oil biodiesel, Figure 16 and pure diesel and ethanol, Figure 17.

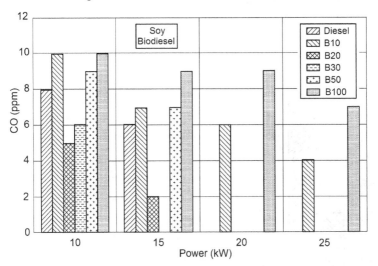

Fig. 15. Concentrations of CO in gas versus the micro-turbine power for different fuels: soy biodiesel.

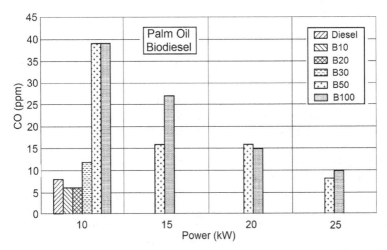

Fig. 16. Concentrations of CO in gas versus the micro-turbine power for different fuels: palm oil biodiesel.

It is observed in Figures 15, 16 and 17 that the concentrations of CO in the exhaust of the micro-turbine operating on pure biodiesel (B100) are higher than when operating with diesel. Emissions decrease as much for diesel as for the blends when the power increases. This if explained by the characteristics of combustion in the combustion chamber.

The soy and palm oil biodiesel has a higher viscosity than the pure diesel. The fuel nozzles not modified in the micro-turbine, must have worsened the quality of atomization of biodiesel compared with diesel, generating higher levels of CO in the exhaust gases as consequence of incomplete combustion. CO emissions at part load are larger than at full load. The lower emission occurs to full load.

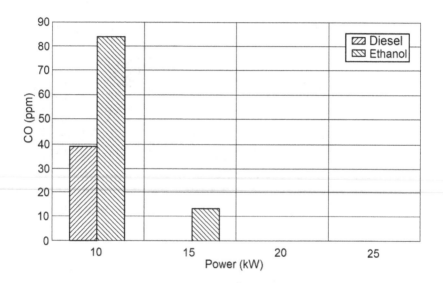

Fig. 17. Concentrations of CO in gas versus the micro-turbine power for different fuels: pure diesel and ethanol.

There are differences in the composition of exhaust gases by using soy and palm oil biodiesel, as can be seen in Figures 15 and 16, as well as been observed previously in thermal efficiency. The operation with soy biodiesel showed no CO in loads exceeding 10 kW, while the operation with palm oil biodiesel (B100) presented CO throughout the power range. The two fuels have different physical-chemical characteristics, which are reflected in a particular behavior in the combustor process.

It is observed in Figure 17 that the emissions of CO to ethanol use are higher than diesel, in all power. This is probably due to lower LHV of ethanol relative to diesel, resulting in higher specific fuel consumption, which reduces the residence time of fuel in the combustion chamber, which may be the cause of greater incomplete combustion.

Figures 18, 19 and 20 show NO_X emissions with different fuels. There are no great differences in the values of NO_X for the fuels.

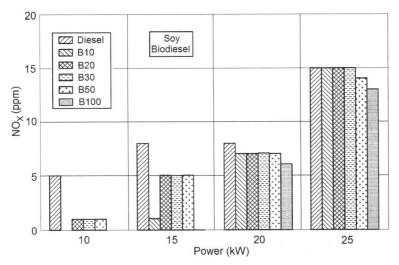

Fig. 18. NO_X emissions versus the micro-turbine power for different fuels: soy biodiesel.

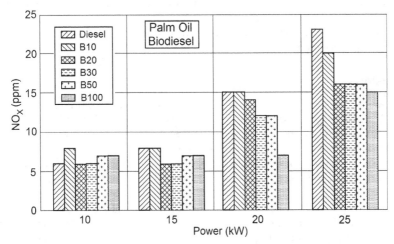

Fig. 19. NO_X emissions versus the micro-turbine power for different fuels: palm oil biodiesel.

NO_X emissions are predominantly of thermal origin and their values are in all cases below 35 ppm, maximum limits set by the engine manufacturer. When used pure biodiesel (B100), the NO_X concentration is lower than the diesel at all loads tested. The results for CO and NO_X show a behavior similar to that presented by PETROV (1999), who also has carried out experiments with a 30 kW micro-turbine.

In the case of ethanol, NO_X emissions showed an inverse behavior to CO, which was expected, because the formation of CO and NO_X is a function of reaction temperature when the CO reduces NO_X increases. Thus, when the amount of CO decreases with increasing power, increases the quantity of NO_X in exhaust gases.

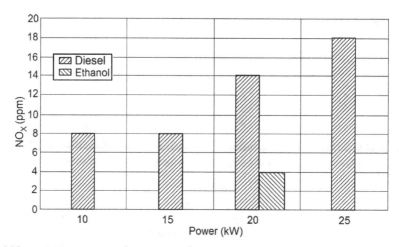

Fig. 20. NO_X emissions versus the micro-turbine power for different fuels: pure diesel and ethanol.

For the three fuels evaluated it was observed a reduction in emissions of NO_X and an increase in CO content compared with diesel, however, all results are within the range indicated by the manufacturer (CAPSTONE, 2001).

It is remarkable that there were no SO_2 emissions and unburned hydrocarbons in any test with biofuels, as mentioned.

6. Conclusions

With increasing industrial development there is the necessity for more refined research on the use of biofuels in gas turbines, covering aspects such as quality of biofuel to be used, form and storage conditions, adjustments in the systems of engines, costs of energy generated in order to maintain high operational reliability of the turbines.

A gas turbine can operate with different types or blends of biofuels with a corrected power loss at around 4.26%, and the corrected heat rate of 8.38% higher than diesel fuel, as shown in this work.

As previously warned care must be taken during the operation of the gas turbine with liquid fuel (from whatever source) and gaseous fuels derived from biomass, because the components of supply systems of gas turbines are very sensitive and the use these biofuels can provide wear or cause loss of efficiency during extended operation.

The physicochemical characteristics of all the fuels evaluated lie within the specifications for their use in gas turbines. The thermal performance tests showed that biodiesel has higher specific consumption than diesel. The reason for that is the lower heat value of the pure biodiesel in comparison with diesel fuels. Agreeing with the findings of other researchers, cited in this work, and verified in tests with biofuels the fuel that presented the smallest difference in terms of heat rate in relation to diesel was the palm oil biodiesel, with a difference of about 17.6 % at full load and less than 1.0 % at medium load.

The tests also found the need to take care with the installations of storage systems and supply of power plants in order to maintain the specifications and properties of mixtures of liquid biofuels within acceptable standards. In the case of gaseous fuels to be careful with he filter system of gas particles and elimination of harmful substances such as tar and others that, besides causing wear to components of the supply system, undermine the performance of engines. Emission levels from the experimental tests have shown that CO increases for the palm oil biodiesel and NO_X decreases by approximately 26.6%; SO_X concentration wasn't taken into account when biodiesel was used. Further investigation involving emission has to be carried out for the better understanding of pollutant formation when biodiesel fuels are used.

7. Acknowledgements

We wish to thank CAPES, FAPEMIG, FAPEPE and CNPq, for their financial support.

8. References

ABNT, NBR15431: "Biodiesel – Determinação de glicerina livre em biodiesel de mamona por cromatografia em fase gasosa", 2006.

ABNT, NBR15432: "Biodiesel – Determinação de monoglicerídeos, diglicerídeos e ésteres totais em biodiesel de mamona por cromatografia em fase gasosa", 2006.

ABNT, NBR15433: "Biodiesel – Determinação da concentração de metanol e/ou etanol por cromatografia em fase gasosa", 2006.

ABNT, NBR15434: "Biodiesel – Determinação de glicerina total e do teor de triglicerídeos em biodiesel de mamona", 2006.

ASME performance test code PTC-22-1997. "Gas turbine power plants".

Biasi, V. "Low cost and High efficiency make 30 to 80 kW microturbines attractive", Gas Turbine World, jan.-fev., Southport, 1998.

Bist, S. "Development of Vegetable Lipids Derived Fatty Acid Methyl Esters as Aviation Turbine Fuel Extenders". Master Thesis of Purdue University, 2004.

Bolszo, C. D.; McDonell, V. G. "Emissions Optmization of a Biodiesel Fired Gas Turbine", Proceedings of the Combustion Institute, 32, Science Direct ELSEVIER, 2949-2956 pp., 2009.

Boyce, P. M. "Gas Turbine Engineering Handbook", Third Edition, Gulf Professional Publishing; april 28, 2006.

Capstone Turbine Corporation. "Capstone Microturbine Model 330 system operation manual", Capstone Turbine Corporation, USA, 2001.

Cohen, H.; Rogers, G. F. C.; Saravanamuttoo, H. I. H., Gas Turbine Theory, Fourth edition, 1996.

Correia, P. S. "The use of biodiesel in gas fuel micro turbine: thermal perforamnce and emission testing". Master Theses, UNIFEI, MG, Brazil, 2006.

Gökalp, I.; Lebas, E. "Alternative Fuels for Industrial Gas Turbines (AFTUR)", Applied Thermal Engineering, 24, 2004.

Habib, Z.; Parthasarathy, R.; Gollahalli, S. "Performance and Emission Characteristcs of Biofuel in a Small-Scale Gas Turbine Engine", Applied Energy 87, 24, 1701-179 pp., 2010.

Kehlhofer, R.; Hannemann, F.; Stirnimann, F.; Rukes, B. "Combined-Cycle Gas & Steam Turbine Power Plants", 3rd Edition, PennWell, 2009, 64-69 pp.

Liss, E. W. "Natural Gas Power System For the Distributed Generation Market", Power-Gen International 99 Conference, New Orleans, Louisiana, 1999.

Lopp, D.; Tanley, D.; Ropp, T.; Cholis, J. "Soy-Diesel Blends Use in Aviation Turbine Engines"; Aviation Technology Department of Purdue University, 1995.

Petrov, A. Y.; Zaltash, A.; Rizy, D. T.; Labinov, S. D. "Study of Flue Gas Emissions of gas Microturbine-Based CHP System"; 1999. www.uschpa.org.

Mimura, N. "Biodiesel Fuel: A next Microturbine Challenge"; 2003. www.ornl.gov.

Nwafor, Omi, 2004. "Emission characteristics of Diesel engine operating on rapeseed methyl ester". Renewable Energy, 29:119-29.

Rodgers, G.; Saravanamutto, H. "Gas Turbine Theory", Editora Prentice Hall, março, 2001.

Schmellekamp, Y.; Dielmann, K. "Rapeseed oil in a Capstone C30"; Workshop: Bio-fuelled Micro Gas Turbines in Europe, Belgium, 2004. www.bioturbine.org.

Schmitz, W.; Hein, D. "Concepts for the Production of Biomass Derived Fuel Gases for Gas Turbine Applications", Proceedings of ASME Turbo Expo 2000, Munich, Germany, may 8-11, 2000.

Scott, W. G. "Micro Gas Turbine Cogeneration Applications", International Power and Light Co., USA, 2000.

Sierra, R. G. A. "Experimental test and thermal economical analysis of biofuel used in regenerative micro gas turbine engine"; Master Theses, UNIFEI, MG, Brazil, 2008.

Tan, E. S., Palanisamy, K., 2008. "Experimental and Simulation Study of Biodiesel combustion in a Micro turbine", ASME Turbo Expo 2008, ASME GT2008-51497.

Wendig, D. "Bio fuel in micro gas turbines", Workshop: Bio-Fuelled Micro Gas Turbines in Europe, Belgium, 2004. www.bioturbine.org.

Willis, H. L.; Scott, W. G. "Distributed Power Generation". Planning and Evaluation, Ed. Marcel Dekker, Inc.

5

Exergy Analysis of a Novel SOFC Hybrid System with Zero-CO$_2$ Emission

Liqiang Duan, Xiaoyuan Zhang and Yongping Yang
School of Energy, Power and Mechanical Engineering,
Beijing Key Lab of Energy Safety and Clean Utilization,
Key Laboratory of Condition Monitoring and Control for Power
Plant Equipment of Ministry of Education,
North China Electric Power University, Beijing,
People Republic of China

1. Introduction

Now, climate change due to the emission of greenhouse gases, especially the emission of CO$_2$, is becoming more and more serious. Though many countries have taken all kinds of measures to control and reduce the emission of CO$_2$, in the short term, CO$_2$ emission still maintains a rapid growth trend. Power industry is the biggest CO$_2$ emission sector. So, there exists the greatest CO$_2$ emission reduction potential in the power industry. Now, many kinds of fossil fuel power generation systems with CO$_2$ recovery are usually based on the chemical absorption method or the oxygen combustion method. The former demands a chemical absorption and separation unit to recover CO$_2$ from the flue gas of power systems. The latter demands a special oxygen combustion technology, equipment and a larger ASU (air separation unit). And these technologies all consume great energy and result in the huger equipment investment and higher operating cost. Now, people are eager to develop the high-efficiency power generation technology with the less energy consumption for CO$_2$ capture. Fuel cell can satisfy the above requirements, with the higher energy conversion efficiency and less energy consumption of CO$_2$ capture, so it has attracted considerable interest in recent years.

Solid Oxide Fuel Cell (SOFC) is an attractive power-generation technology that can convert the chemical energy of fuel directly into electricity while causing little pollution (Kartha & Grimes, 1994). Because the anode fuel gas is naturally separated from the cathode air by the solid electrolyte, the CO$_2$ gas with the higher concentration can be obtained in the anode exhaust gas. In addition, SOFC can employ all kinds of fuels, including various hydrocarbon fuels. Compared with the traditional power generation systems, the SOFC hybrid system power plant has the higher system efficiency (net AC/LHV). Even after CO$_2$ is captured, the efficiency of SOFC hybrid system still can be greater than or equal to that of the traditional power systems without CO$_2$ capture. In order to further improve the CO$_2$ concentration of anode exhaust gas, SOFC can employ the O$_2$/CO$_2$ combustion mode in the afterburner. Because the required mass flow of pure O$_2$ is less, the energy consumption is lower. After capturing the CO$_2$, the SOFC hybrid system does not result in a bigger efficiency reduce. So the SOFC hybrid power system with zero CO$_2$ emission become a new way which can simultaneously solve the problem of efficient energy utilization and lower pollution emission.

In the last decades, many researchers were involved in study of SOFC stack and the hybrid power system with CO_2 capture. Y.Inui proposed and investigated two types of carbon dioxide recovering SOFC/GT combined power generation systems in which a gas turbine with carbon dioxide recycle or water vapor injection is adopted at the bottoming cycle system (Y.Inui et al, 2005). The overall efficiency of the system with carbon dioxide recycle reaches 63.87% (HHV) or 70.88% (LHV), and that of the system with water vapor injection reaches 65% (HHV) or 72.13% (LHV). A. Franzoni considered two different technologies for the same base system to obtain a low CO_2 emission plant (Franzoni et al, 2008). The first technology employed a fuel decarbonization and CO_2 separation process placed before the system feed, while the second integrated the CO_2 separation and the energy cycle. The result showed that the thermodynamic and economic impact of the adoption of zero emission cycle layouts based on hybrid systems was relevant. Philippe Mathieu presented the integration of a solid oxide fuel cell operating at a high temperature (900°C–1000°C, 55–60% efficiency) in a near-zero emission CO_2/O_2 cycle (Philippe Mathieu, 2004). Takeshi Kuramochi compared and evaluated the techno-economic performance of CO_2 capture from industrial SOFC-combined heat and power plant (CHP) (Takeshi et al, 2009). CO_2 is captured by using an oxyfuel afterburner and conventional air separation technology. The results were compared to both SOFC-CHP plants without CO_2 capture and conventional gas engines CHP without CO_2 capture. B.Fredriksson Moller examined the SOFC/GT configuration with and without a tail-end CO_2 separation plant, and based on a genetic algorithm, selected the key parameters of the hybirid system (Fredriksson et al, 2004). The result of the optimization procedure shows that the SOFC/GT system with part capture of the CO_2 exhibits an electrical efficiency above 60%. Some researchers also studied the performance parameters of the different SOFC hybrid power systems from the thermoeconomic or exergy efficiency point (Bozzolo et al, 2003; Asle & Matteo 2001; Takuto et al, 2007). For example, Ali Volkan Akkaya proposed a new criterion-exergetic performance coefficient (EPC), then applied it in the SOFC stack and SOFC/GT CHP system (Ali et al, 2007, 2009). F. Calisa discussed the simulation and exergy analysis of a hybrid SOFC-GT power system. The result showed that the SOFC stack was the most important sources of exergy destruction (Calisea et al, 2006).

In this paper, a zero-CO_2 emission SOFC hybrid power system is proposed. Using exergy analysis method, the exergy loss distributions of every unit of zero-CO_2 emission SOFC hybrid system are revealed. The effects of different operating parameters on exergy loss of every unit, as well as the overall system performance, are also investigated. The results obtained in this paper will provide useful reference for further study on high-efficient zero emission CO_2 power system.

2. System modelling

The models developed in the paper are all based on the following general assumptions:

1. All components work in adiabatic conditions, pressure drops and refrigerant disclosure are all neglected, and the systems operate at steady-state conditions.
2. The cathode gas consists of 79% nitrogen and 21% oxygen, and all gases are assumed as ideal gases.

3. The mass flow of the input fuel, gas and all the reaction products are stable, the changes of the fluid kinetic energy and potential energy are neglected.
4. The unreacted gases are assumed to be fully oxidized in the after-burner of the SOFC stack, and the after-burner is assumed to be insulation, all the heat exchangers are adiabatic.
5. The temperature of the anode and cathode outlet gases are equal to the cell stack operating temperature, the current and voltage of every cell unit are the same.

2.1 The SOFC stack model and result analysis
2.1.1 SOFC model
The natural gas feed tubular SOFC system process is implemented by using Aspen Plus software. The Aspen Plus contains rigorous thermodynamic and physical property database and provides comprehensive built-in process models, thus offering a convenient and time saving means for chemical process studies, including system modeling, integration and optimization. The simulated SOFC flowsheet is shown in Figure 1. It includes all the components and functions contained in the SOFC stack, such as ejector, pre-reformer, fuel cell (anode and cathode) and afterburner.

Firstly, the preheated fuel (stream 1) mixes with the recycling anode exhausted gas (stream 6), and then the mixed fuel gas (stream 2) is sent to the pre-reformer where the steam reform reaction takes place. After that, the stream (4) enters the anode of SOFC in which the electrochemical reaction of fuel and oxygen from the anode occurs. The reaction product and unreacted flue mixture (stream 5) is separated into two parts. One part (stream 6) is recycled. Another part enters the afterburner and mixes with the nitrogen-rich air (stream 13) from the anode. After the combustion reaction, the exhausted gas from the afterburner (stream 14) is introduced into the regenerator to preheat the air (stream 9) for the anode.

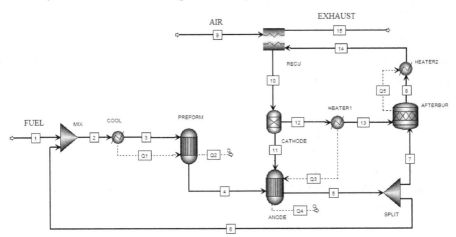

Fig. 1. Aspen Plus SOFC Stack Model Flowsheet

The cell voltage calculation is the core of any fuel cell modeling. The semi-empirical equations from literature (Stefano, 2001) were used to compose the Aspen Plus calculation module to simulate these effects on voltage. Several Design-spec Fortran blocks are used to set the fuel cell system's energy and heat balance. The semi-empirical equations are as follows:

$$\Delta V_p(mV) = C_1 \log(\frac{p}{p_{ref}}) \tag{1}$$

$$\Delta V_T(mV) = K_T(T - T_{ref}) \times i_c \tag{2}$$

$$\Delta V_{an}(mV) = 172 \log \frac{p_{H_2}/p_{H_2O}}{(p_{H_2}/p_{H_2O})_{ref}} \tag{3}$$

$$\Delta V_{cat}(mV) = 92 \log(\frac{p_{O_2}}{(p_{O_2})_{ref}}) \tag{4}$$

By summing the above four correlations, the actual voltage V can be calculated as

$$V_c = V_{ref} + \Delta V_p + \Delta V_T + \Delta V_{cat} + \Delta V_{an} \tag{5}$$

Where i_c is the current density; p_{H_2}/p_{H_2O} is the ratio of H_2 and steam partial pressure; P_{O_2} is the average oxygen partial pressure at the cathode for the actual case; V_c is the actual cell voltage; the subscribe ref is the reference case. P_{ref} = 1 bar; T_{ref} = 1000°C; $(p_{O_2})_{ref}$ = 0.164; $(p_{H_2}/p_{H_2O})_{ref}$ =0.15. ΔV_P stands for the change of SOFC voltage when the operating pressure varies. C_1 is the theoretical Nernst constant. ΔV_T stands for the change of SOFC voltage when the operating temperature varies. K_T is equal to 0.008 when the operating temperature is in the range of 900-1050°C. ΔV_{cat} and ΔV_{an} stand for the voltage change when the cathodic flow composition and anode flow composition vary compared with the reference case, respectively.

The fuel cell power output is the product of the cell voltage and current.

$$Power = current(A) \times V(V) \tag{6}$$

The equivalent hydrogen flow rate $n_{H_2,equivalent}$, can be calculated based on the molar flow rate of H_2 consumed in the electrochemical reaction ($n_{H_2,consumed}$) and the fuel utilization factor (U_f) to be generated:

$$n_{H_2,equivalent}(mol/h) = \frac{n_{H_2,consumed}}{U_f} \tag{7}$$

The value of the equivalent hydrogen flow rate and the known inlet fuel composition (C_i) can determine the amount of fresh fuel required $n_{fresh,fuel}$ from the following equation (8):

$$n_{fresh,fuel}(mol/h) = \frac{n_{H_2,equivalent}(mol/h)}{C_{H_2} + C_{CO} + 4 \times C_{CH_4} + 7 \times C_{C_2H_6} + \cdots} \tag{8}$$

The cell electrical efficiency is calculated according to the following equation (9):

$$\eta = \frac{Power}{n_{fresh,fuel}(mol/h) \times LHV_{fuel}(J/mol)} \tag{9}$$

2.1.2 SOFC simulation result analysis

In the process of modeling the hybrid power system, the accuracy of the fuel cell stack simulation is critical for the overall system. So this paper firstly checked the accuracy of SOFC stack model. With the same input parameters of the literatures (W.Zhang et al, 2005), the simulation results were compared with those of the literatures (W.Zhang et al, 2005,Veyo, 1996,1998,1999) . As shown in Table 1, the results show that this paper's simulation results are very close to the literature results. So the model of SOFC stack is feasible and reliable, it can be applied to simulate the overall SOFC hybrid system.

Parameters (unit)	Literature Data (W,Zhang et al 2005)	Literature Data (Veyo, 1996,1998, 1999)	Simulation data of this paper
Voltage (V)	0.70	-	0.70
Current Density (mA/cm^2)	178	180	179.1
Air utilization factor	19%	-	18.2%
Pre-reformer Outlet Temperature (℃)	536	550	537.2
Stack Exhaust Composition (EXHAUST)	77.3%N$_2$ 15.9%O$_2$ 4.5%H$_2$O 2.3%CO$_2$	77%N$_2$ 16%O$_2$ 5%H$_2$O 2%CO$_2$	77.5%N$_2$ 16.4%O$_2$ 4%H$_2$O 2.1%CO$_2$
Anode Outlet Composition (stream5)	50.9% H$_2$O 24.9%CO$_2$ 11.6%H$_2$ 7.4%CO 5.1%N$_2$	48% H$_2$O 28%CO$_2$ 14%H$_2$ 5%CO 5%N$_2$	50.9% H$_2$O 24.9% CO$_2$ 11.6% H$_2$ 7.4%CO 5.1%N$_2$
Stack Exhaust Temperature (℃)	834	847	832
Stack Efficiency (AC)	52%	50%	51.99%

Table 1. The Simulation Results of SOFC Stack （120kW DC output）

2.2 Description of the basic SOFC hybrid system

The basic SOFC hybrid system uses a tubular SOFC stack with higher operating temperature. Figure 2 shows the flowsheet of the system. The fuel is compressed and preheated, then is put into pre-reformer to generate the required H$_2$, CO and CO$_2$. Air is supplied by a blower and preheated prior to enter the SOFC stack. Then air participates an electrochemical reaction with fuel in the fuel cell stack. Because the pre-reformer needs a larger amount of water vapor, a hot recycle stream from the anode outlet is directed to the pre-reformer. The outlet stream of SOFC anode is mainly composed of H$_2$O, CO$_2$ and unconverted fuel (H$_2$ and CO). Part of this stream is injected into after-burner, then the hot outlet gas with high pressure expands in gas turbine, the rest is recycled and mixed with the compressed and preheated fuel. The recirculation fraction is calculated to obtain a given steam/carbon (S/C) ratio. In this way, the system can prevent from carbon deposition phenomenon, enhance the pre-reformer temperature and get more H$_2$. Finally, the product gas of SOFC is sent into the after-burner, where the unreacted fuel is burnt with part of the excess air from the cathode. After expansion of the hot gas in gas turbine, it is sent into heat exchangers to heat the inputted fuel and gas.

In order to control the outlet gas temperature and prevent the temperature from exceeding the tolerable inlet temperature of the turbine, water vapor is injected into the afterburner.

After the hot outlet gas with high pressure expands in gas turbine, it is fed to recuperator to heat the inputted fuel and air. Finally it is fed to the heat exchanger 4 for producing the saturated water vapor required for the after-burner.

Although SOFC hybrid power system has realized energy conversion with higher efficiency, the hybrid system still emit some greenhouse gas, which contains a large number of N_2 that will greatly impact the energy consumption of capturing CO_2. The paper proposed a zero CO_2 emission SOFC hybrid power system with water vapor injection and made the detailed exergy analysis on it.

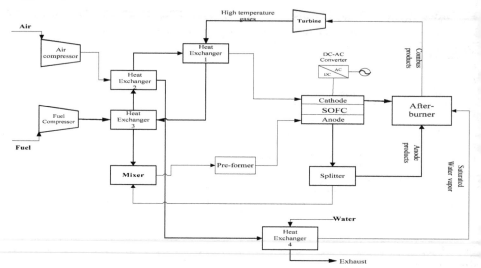

Fig. 2. Flowsheet of the basic SOFC hybrid system

2.3 Description of the new SOFC hybrid system with CO_2 capture

The new system and the basic system's primary processes are the same, but the new system employs O_2/CO_2 combustion mode. Pure oxygen from air separation unit (ASU) is fed into the combustion chamber to burn with the anode exhaust gas. The product gas only contains water vapor and CO_2. As shown in Figure 3, the compressed and preheated air is supplied to enter the SOFC stack and participate in an electrochemical reaction with fuel in the SOFC stack. At the same time, fuel is supplied by a blower and preheated prior to entering the SOFC stack anode. Part of the anode product gas is recycled to the mixer, and then mixed with the input compressed and preheated air; the remaining stream is burnt in the after-burner using oxygen as the oxidizer. After the combustion products expand in gas turbine, they are fed to heat exchangers to heat the inputted fuel and air. Finally the product gases are injected into the heat exchanger 4 and preheat the feed water for the afterburner. Theoretically the produced steam is enough to reduce the combustion gas temperature to the acceptable inlet temperature of turbine. The obtained combustion gas, which is composed of only CO_2 and water vapor, is introduced to gas turbine. After expansion, the cathode products are also injected into the heat exchanger 4 (also called the fourth heat exchange). Generally, capturing CO_2 is a difficult and complicated process, but the system uses O_2/CO_2 combustion mode, as a result, the combustion product gases only contain water vapor and CO_2. CO_2 can be separated by condensation, which does not consume extra energy. Finally it is fed to the multi-stage compressor and makes CO$_2$ dried from water and liquefied for making the transportation easier.

In order to realize zero CO$_2$ emission and lower the energy consumption of CO$_2$ capture, the new system mainly adopts the following measures:

1. In order to prevent the dilution of N$_2$, only the anode exhaust gas is burned with pure oxygen in the after-burner.
2. In order to enhance the system power, cathode outlet gas of SOFC stack is channeled into turbine to expand. And the product gas is channeled into the heat exchanger 4 to produce enough water vapor for the afterburner.
3. To avoid the great energy consumption, the new system uses multi-stage compression mode. After removing H$_2$O, the CO$_2$ concentration of depleted gas is more than 99%.

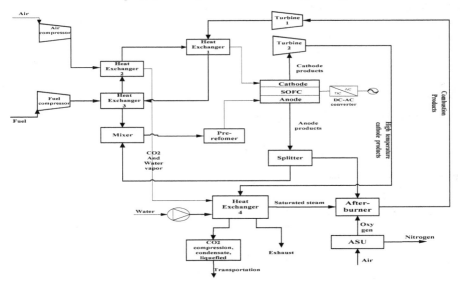

Fig. 3. Flowsheet of the zero CO$_2$ emission SOFC hybrid system

2.4 Simulation analysis of the basic and the new system

(1) System simulation parameters

The inputted fuel compositions of the basic and the new systems are the same (CH$_4$ 93.6%, C$_2$H$_6$ 4.9%, C$_3$H$_8$ 0.4%, C$_4$H$_{10}$ 0.2%, CO 0.9%). The fuel utilization factor is 85%, and the air utilization rate is 25%. The SOFC operating temperature/pressure is 1123.15K/3atm. The steam turbine isentropic efficiency is 85% and the gas turbine isentropic efficiency is 80%. The DC to AC inverter efficiency is 92%. The efficiencies of CO$_2$ compressor and air compressor in ASU are 75%.

(2) The performance analysis of the two systems

Table 2 shows the simulation results of the basic and new systems. For the new system, the cathode gas of the new system has a certain pressure; the new system can recover some expansion work of cathode exhaust gas. Though the new system applies the air separation unit and CO$_2$ capture equipment, which will consume some energy, compared with the cathode gas expansion work, it can be neglected.

Parameters (unit)	Base case	New system
SOFC Voltage (V)	0.64	0.64
SOFC Current Density (mA/cm²)	180.54	180.54
SOFC Stack Efficiency (%)	47.98	47.98
Electrical Efficiency (%) (AC)	59.03	56.26

Table 2. The simulation results of the two systems

3. The exergy analysis of the new hybrid system with CO_2 capture

Exergy analysis method is based on the second thermodynamics Law, considering both the energy quantity and energy quality in every energy conversion process. On the base of the energy balance and mass balance, the overall system is divided into many units and the exergy value of every stream is calculated. Through exergy balance analysis, the distributions of exergy loss of every unit can be acquired. At the same time, the system's largest exergy loss unit can be found. So, the location and direction of reducing exergy loss to improve the efficiency of the system can be identified (Jiaxuan & Shufang 1993).

The total exergy loss ($E_{X_{D,tot}}$) of the hybrid power system is given by the following equation:

$$E_{X_{D,tot}} = E_{X_{D,SOFC}} + E_{X_{D,GT}} + E_{X_{D,ASU}} + \ldots\ldots\ldots\ldots \quad (10)$$

Where $E_{X_{D,SOFC}}$ is SOFC stack's exergy loss; $E_{X_{D,GT}}$ is the turbine's exergy loss; $E_{X_{D,ASU}}$ is the ASU's exergy loss;

In order to analyze the system exergy loss in detail, the exergy loss rate (σ_i) is designed as the ratio of every component exergy loss ($E_{X_{D,i}}$) to the total system exergy loss.

$$\sigma_i = \frac{E_{X_{D,i}}}{E_{X_{D,tot}}} \times 100\% \quad (11)$$

The exergy efficiency (η_E) of the SOFC hybrid power system is another important exergetic performance criterion, and it can be defined as the ratio of total system effective acquired exergy to fuel input exergy as the following equation:

$$\eta_E = \frac{W + E_Q}{E_F} \times 100\% \quad (12)$$

Where, W is the system power; E_Q is the acquired heat exergy; E_F is the fuel exergy, and it can be calculated as the following equation:

$$E_F = Q_{DW} \times U_{fuel} \quad (13)$$

Where, Q_{DW} is the lower heating value (LHV) of the input fuel; U_{fuel} is the mass flow rate of fuel.

Exergy analysis takes the environmental state as reference state (P_0, T_0). In this paper, the selected temperature/pressure is 1123.15K/1atm as the reference environment state.

3.1 Exergy loss analysis of the system's every unit

Figure 4 shows the every unit exergy loss of the zero CO_2 emission SOFC hybrid power system. The biggest exergy loss unit lies in SOFC stack, which accounts for more than 35% of the total exergy loss. The main reason is that excess air is injected into to the SOFC stack in order to reduce the temperature difference of SOFC stack, and part of the input fuel chemical energy heats the excess air, which will cause a significant irreversible loss. One part of energy generated by electrochemical reaction is directly converted into the electrical energy, while the other part is changed into heat power to ensure that the fuel is reformed into H_2. So it makes the useful work generated by the fuel chemical energy reduce and the exergy loss increase.

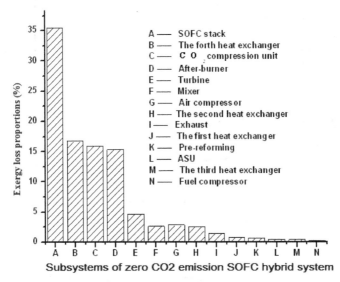

Fig. 4. Exergy loss distributions of Zero CO_2 emission hybrid power system units

According to the Second Law of Thermodynamics, even the heat loss of heat exchanger is neglected, there is still irreversible exergy loss in the inside of heater caused by big temperature difference heat transfer and mucous membrane resistance in the flow process of cold and hot fluid (Calise et al, 2006). As shown in Figure 4, the exergy loss of the fourth heat exchanger is the second biggest. In order to effectively reduce the exergy loss of heat exchangers, the heat transfer process should be designed reasonably in order to reduce the temperature difference.

3.2 Parametric exergy analysis results and discussions

The operating temperature, the operating pressure, the current density and fuel utilization factor of SOFC system are all considered as key variables which greatly influence the overall system performance. In the following discussion, the effects of the above key variables on the exergetic performance of system are respectively discussed.

3.2.1 The operating temperature

When the mass flow of input fuel keeps constant, with the increase of the operating temperature of SOFC, both the fuel cell voltage and system exergy efficiency increase. And then, the required air for cooling fuel cell stack will decrease as shown in Figure 5. In addition, due to the enhancement of cell stack activity, the system exergy loss reduces and the total system output power increases as shown in Figure 6. When the operating temperature is above 920°C, the voltage begins to decrease and system exergy losses increased. Therefore, in the practical situation, the system should operate in the proper temperature.

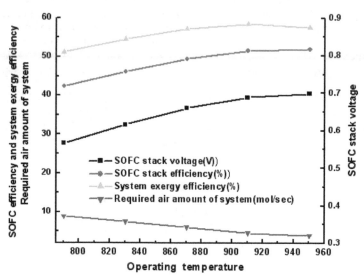

Fig. 5. The effect of operating temperature on system performance

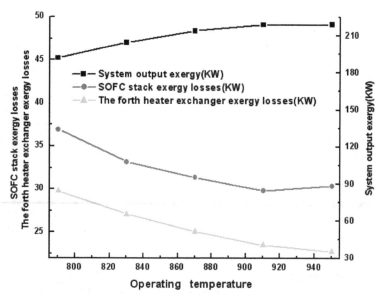

Fig. 6. The effect of operating temperature on system exergy parameters

3.2.2 The operating pressure

The operating pressure is vital to the system performance. Improving the operating pressure of SOFC stack, the SOFC voltage will increase because the H$_2$ amount in SOFC stack and H$_2$ partial pressure increase. Figure 7 shows that keeping the current density constant, with the increase of the operating pressure, the voltage increases. However, the growth rate gets smaller.

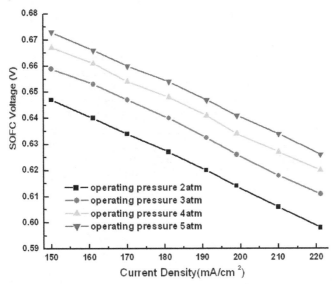

Fig. 7. The effect of operating pressure on SOFC Voltage

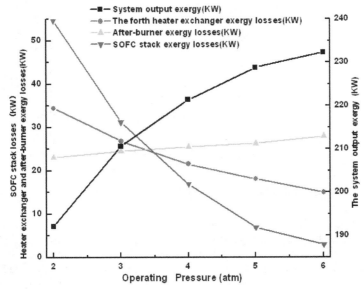

Fig. 8. The effect of operating pressure on system exergy performance

As shown in Figure 8, as the operating pressure increases, the SOFC stack exergy loss decreases and the total system output exergy increases. Because the required air for cooling fuel cell stack slowly increases, the after-burner exergy loss increases and the exergy loss of heat exchanger 4 decreases. In a word, the higher operating pressure is favorable to improving the performance of SOFC hybrid system. However, the higher pressure will increase the cost of system investment. Choosing the appropriate operating pressure should be taken into account when designing the SOFC.

3.2.3 The fuel utilization factor (U_f)

The fuel utilization factor (U_f) has a significant effect on the cell voltage and efficiency. As shown in Figure 9, with the increase of U_f from 0.7 to 0.9, the current density will increase, which will result in the decrease of the cell voltage. At lower values of U_f, when U_f increases, the cell voltage change is not significant, so the system output exergy will increase (as shown in Figure 10). But for higher U_f, the change amount of the cell voltage is bigger than that of the current density, as a result, the system exergy efficiency will reduce as shown in Figure 8. And U_f also has a significant impact on the composition of the anode exhaust stream. The CO_2 concentration at the anode outlet increases when U_f is increased because the fuel is more depleted (less CO and H_2), which will result in the change of the system unit exergy loss as shown in Figure 10.

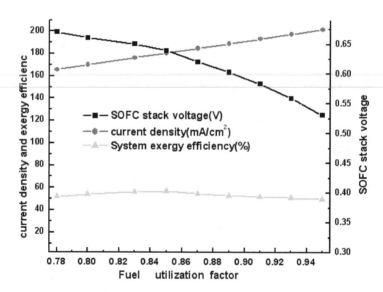

Fig. 9. The effect of fuel utilization factor on system performance

3.2.4 The cathode air input temperature

The cathode air consists of 79% nitrogen and 21% oxygen. As shown in Figure 11, with the increase of the cathode air input temperature, the activity of SOFC stack enhances. At the same time, both the required air for cooling fuel cell stack and the SOFC voltage increase, as a result, the system will produce more power. In addition, the inlet turbine gas temperature also increases, the power output of turbine will boost. But in order to meet the requirement

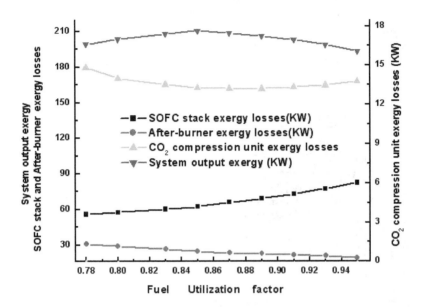

Fig. 10. The effect of fuel utilization factor on system exergy performance

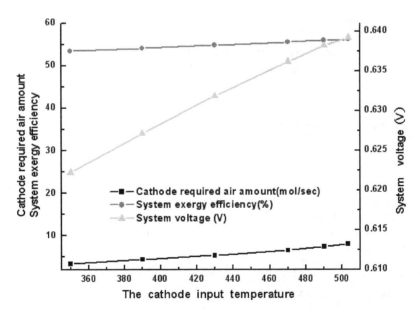

Fig. 11. The effect of cathode air input temperature on system performance

of the inlet air temperature, more heat of the exhaust gas will be consumed. The corresponding exergy loss of heat exchanger increases, so the system exergy efficiency isn't significant increased as shown in Figure 11. And as shown in Figure 12, the input temperature of cathode air also has an important effect on the other system performance parameters. The lower temperature will make the SOFC stack performance deteriorate.

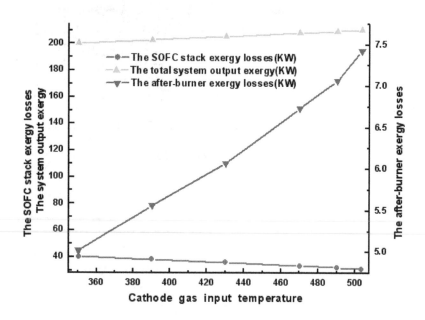

Fig. 12. The effect of cathode air input temperature on system exergy performance parameters

3.2.5 The oxygen concentration effect

As can be seen from Figure 13, when the operating pressure is a constant, as the oxygen purity increases, the O_2 partial pressure of SOFC stack cathode air improves, and this will make the system output exergy and exergy efficiency increase, especially SOFC stack with the lower operating pressure. Because the fuel flow remains unchanged, with the increase of oxygenconcentration, the required air decreases. Due to that the electrochemical reaction is exothermic reaction, it may cause the local area of stack overheat and the battery performance deteriorate. And with the increase of the oxygen concentration the consumed energy for separating the air will become bigger, so the exergy loss of the stack will rise slowly, which will result in the slow rise tend of system output exergy (as shown in Figure 14).

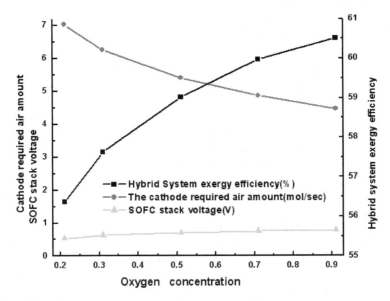

Fig. 13. The effect of oxygen concentration on system performance

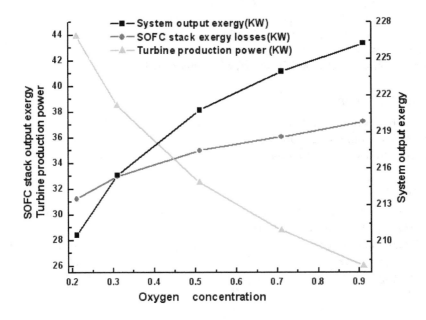

Fig. 14. The effect of oxygen concentration on system exergy performance

4. Conclusions

Based on a traditional SOFC (Solid Oxide Fuel Cell) hybrid power system, A SOFC hybrid power system with zero-CO_2 emission is proposed in this paper and its performance is analyzed. The exhaust gas from the anode of SOFC is burned with pure oxygen and the concentration of CO_2 gas is greatly increased. Because the combustion produce gas is only composed of CO_2 and H_2O, the separation of CO_2 hardly consume any energy. At the same time, in order to maintain the proper turbine inlet temperature, the steam produced from the waste heat boiler is injected into the afterburner, and then the efficiency of hybrid power system is greatly increased. With the exergy analysis method, this paper studied the exergy loss distribution of every unit of SOFC hybrid system with CO_2 capture and revealed the largest exergy loss unit. The effects of main operating parameters on the overall SOFC hybrid system with CO_2 capture are also investigated.

The research results show that the new zero-CO_2 emission SOFC hybrid system still has a higher efficiency. Its efficiency only decreases 3 percentage points compared with the basic SOFC hybrid system without CO_2 capture. The O_2/CO_2 combustion mode can fully burn the anode's fuel gas, and increase the concentration of CO_2 gas; at the same time with the steam injection and the combustion products are channeled into turbine, the efficiency of system greatly increases. The liquefaction of CO_2 by the mode of multi-stage compression and intermediate cooling can also greatly reduce the energy consumption.

The exergy analysis of the zero CO_2 emission SOFC hybrid power system shows that SOFC stack, after-burner and CO_2 compression unit are the bigger exergy loss components. By improving the input temperature of SOFC stack and turbine, the system exergy loss will significantly reduce. The optimal values of the operation parameters, such as operating pressure, operating temperature and fuel utilization factor exist, which make the system efficiency highest. Above research achievements will provide the new idea and method for further study on zero emission CO_2 system with higher efficiency.

5. Acknowledgments

This study has been supported by the National Basic Research Program of China (No. 2009CB219801), National Nature Science Foundation Project (No.50606010), and "the Fundamental Research Funds for the Central Universities" (No.10ZG03)

6. References

Kartha S, Grimes P (1994). Fuel cells: energy conversion for the next century. Physics Today, Vol 47(11) : 54–61,ISSN:0031-9228.

Y.Inui, T.Matsumae, H.Koga, K.Nishiura (2005). High performance SOFC/GT combined power generation system with CO_2 recovery by oxygen combustion method. Energy Conversion and Management, Vol 46(11-12): 1837–1847, ISSN:0196-8904.

A. Franzoni, L. Magistri, A. Traverso, A.F. Massardo (2008). Thermoeconomic analysis of pressurized hybrid SOFC systems with CO$_2$ separation. Energy, Vol 33(2): 311–320, ISSN: 0360-5442.

Philippe Mathieu (2004). Towards the hydrogen era using near-zero CO$_2$ emissions energy systems. Energy Vol 29(12-15): 1993–2002,ISSN: 0360-5442.

Takeshi Kuramochi, Hao Wu, Andrea Ramirez, Andre Faaij and Wim Turkenburg(2009). Techno-economic prospects for CO$_2$ capture from a Solid Oxide Fuel Cell – Combined Heat and Power plant. Preliminary results. Energy Procedia, Vol. 1(1): 3843–3850, ISSN:1876-6102.

B. Fredriksson Möllera, J. Arriagadaa, M. Assadia, I. Pottsb (2004). Optimisation of an SOFC/GT system with CO$_2$-capture. Journal of Power Sources, Vol. 131(1-2): 320–326, ISSN:0378-7753.

Bozzolo M, Brandani M, Traverso A, Massardo AF(2003). Thermoeconomic analysis of gas turbine plants with fuel decarbonization and carbon dioxide sequestration, ASME transactions. J Eng Gas Turbines Power, Vol. 125(4): 947–53.ISSN:0742-4795.

Asle Lygre, Matteo Ce (2001). Solid Oxide Fuel Cell Power with Integrated CO$_2$ capture. Second Nordic Minisymposium on Carbon Dioxide Capture and Storage, Gothenburg, Sweden.

Takuto Araki, Takuya Taniuchi, Daisuke Sunakawa (2007). Cycle analysis of low and high H$_2$ utilization SOFC/gas turbine combined cycle for CO$_2$ recovery, Journal of Power Sources, Vol 171: 464–470, ISSN:0378-7753.

Ali Volkan Akkaya, Bahri Sahin , Hasan Huseyin Erdem (2009). Thermodynamic model for exergetic performance of a tubular SOFC module. Renewable energy, Vol 34(7):1863-1870, ISSN:0960-1481.

Ali Volkan Akkaya, Bahri Sahin, Hasan Huseyin Erdem (2007). Exergetic performance coefficient analysis of a simple fuel cell system. International Journal of Hydrogen Energy,Vol. 32: 4600– 4609, ISSN:0360-3199.

F. Calisea, M. Dentice d'Accadiaa, A. Palomboa, L. Vanoli (2006). Simulation and exergy analysis of a hybrid Solid Oxide Fuel Cell (SOFC)–Gas Turbine System. Energy, Vol 31 : 3278–3299, ISSN: 0360-5442.

Stefano Campanari (2001).Thermodynamic model and parametric analysis of a tubular SOFC module.Journal of Power Sources, Vol. 92(1-2): 26-34, ISSN:0378-7753.

W. Zhang, E. Croiset, P.L.Douglas, M.W.Fowler, E.Entchev. Simulation of a tubular solid oxide fuel cell stack using AspenPlusTM unit operation models. Energy Conversion and Management 46(2005) 181-196, ISSN: 0196-8904.

Veyo SE. The Westinghouse solid oxide fuel cell program—a status report (1996). In: Proceedings of the 31st IECEC, No. 96570, pp: 1138-43.

Veyo SE, Forbes CA. Demonstractions based on Westinghouse's prototype commercial AES design. In: Proceedings of the Third European Solid Oxide Fuel Cell Forum, 1998, p. 79-86.

Veyo S, Lundberg W. Solid oxide fuel cell power system cycles. ASME Paper 99-GT-356, International Gas Turbine and Aeroengine Congress and Exhibition, Indianapolis, June 1999.

Jiaxuan, Wang, Shufang, Zhang (1993). Exergy method and its application in power plants, China Water Power Press, ISBN: 7-120-01797-7

6

Flexible Micro Gas Turbine Rig for Tests on Advanced Energy Systems

Mario L. Ferrari and Matteo Pascenti
University of Genoa – Thermochemical Power Group (TPG)
Italy

1. Introduction

During the last 15-20 years, microturbine (mGT) technology has become particularly attractive for power generation, especially in the perspective of the development of a distributed generation market (Kolanowski, 2004). The main advantages related to microturbines, in comparison to Diesel engines, are:

- smaller size and weight;
- fuel flexibility;
- lower emissions;
- lower noise;
- vibration-free operation;
- reduced maintenance.

The development of a laboratory based on a large size gas turbine is usually not feasible at the University level for high costs of components and plant management. However, the microturbine (mGT) technology (Kolanowski, 2004) allows wide-ranging experimental activities on small size gas turbine cycles with a strong cost reduction.

Moreover, this technology is promising from co-generative (and tri-generative) application point of view (Boyce, 2010), and is essential for advanced power plants, such as hybrid systems (Massardo et al., 2002), humid cycles (Lindquist et al., 2002), or externally fired cycles (Traverso et al., 2006).

However, if microturbine standard cycle is modified by introducing innovative components, such as fuel cells (Magistri et al., 2002, 2005), saturators (Pedemonte et al., 2007) or new concept heat exchangers (such as ceramic recuperators (McDonald, 2003)), at least two main aspects have to be considered:

- avoiding dangerous conditions (e.g.: machine overspeed, surge, thermal and mechanical stress, carbon deposition);
- ensuring the proper feeding conditions to both standard and additional components.

Moreover, the operation with new devices generates additional variables to be monitored, new risky conditions to be avoided and requires additional control system facilities (Ferrari, 2011).

Experimental support is mandatory to develop advanced power plants based on microturbine technology and to build reliable systems ready for commercial distribution. A possible cheap solution to perform laboratory tests is related to emulator facilities able to generate similar effects of a real system. These are experimental rigs designed to study various critical aspects of advanced power plants without the most expensive components (e.g. fuel cells, high temperature heat exchangers). An important experimental study with mGT based emulators is running at U.S. DOE-NETL laboratories of Morgantown (WV-USA). This activity is based on a test rig designed to emulate cathode side of hybrid systems based on Solid Oxide Fuel Cell (SOFC) technology (Tucker et al., 2009). It is mainly composed of a recuperated microturbine, a fuel cell vessel (without ceramic material), an off-gas burner vessel, and a combustor controlled by a fuel cell real-time model (Tucker et al., 2009). Another emulator facility equipped with a micro gas turbine is under development at German Aerospace Center (DLR), Institute of Combustion Technology, of Stuttgart (Germany) (Hohloch et al., 2008). Its general layout is similar to the NETL test rig. However, this experimental plant includes a fuel cell vessel able to emulate the real stack exhaust gas composition with a water cooling system (coupled with an additional combustor) (Hohloch et al., 2008).

At University of Genoa, TPG researchers developed a new test rig based on a commercial recuperated 100 kW micro gas turbine equipped with a set of additional pipes and valves for flow management. These pipes are essential to perform high fidelity mass flow rate measurements and to connect the machine to an external modular vessel. This additional volume is located between the recuperator outlet (cold side) and the combustor inlet (Ferrari et al., 2009a) to emulate the dimensions of advanced cycle components (such as fuel cells (Ferrari et al., 2009a; Tucker et al., 2009), externally fired gas turbine facilities (Traverso et al., 2006), saturators (Pedemonte et al., 2007)).

This test rig, developed to carry out experimental tests in both steady-state and transient conditions, is designed to have the highest plant flexibility performance. In details, this facility is able to operate in the following configurations:

- simple cycle;
- recuperated and partly recuperated cycle;
- both simple and recuperated cycles coupled with a modular vessel for the emulation of additional component volume (Ferrari et al., 2009a);
- emulation of hybrid systems (Ferrari et al., 2010a) with high temperature fuel cell stack (cathodic and anodic vessels, anodic recirculation (Ferrari et al., 2010a), steam injection for chemical composition emulation, real-time model for components not present in the rig);
- co-generative and tri-generative (with an absorber cooler) systems.

Moreover, on all these plant layouts it is possible to test the influence of the following properties, especially in transient conditions:

- ambient temperature;
- volume size (downstream of the compressor);
- valve fractional opening values;
- bleed mass flow rates;
- grid connection or stand-alone systems;
- control system.

This chapter shows some examples of tests carried out with the rig (using different plant layouts and operative conditions), inside different research projects and during educational activities. In details, the facility was mainly developed inside two Integrated Projects of the EU VI Framework Program (Felicitas and Large-SOFC) and now it is involved in the new EU VII Framework (E-HUB Project) for tests to be carried out with an absorption cooler (tri-generative configuration). Moreover, this experimental plant is essential to introduce undergraduate students to micro gas turbine technology, and Ph.D.s to advanced experimental activities in the same field. With this experimental rig, in addition to learning about thermodynamic cycles and plant layouts, students can also become familiar with their materials, piping, gaskets, technology for auxiliaries, and instrumentation.

2. Nomenclature

CAD	Computer Aided Design
CC	Combustion Chamber
CFD	Computational Fluid Dynamic
Ex	heat Exchanger
FC	Fuel Cell
GT	Gas Turbine
HS	Hybrid System
ISO	International Organization for Standardization
LAN	Local Area Network
mGT	micro Gas Turbine
mHAT	micro HAT cycle
NETL	National Energy Technology Laboratory
REC	RECuperator
RRFCS	Rolls-Royce Fuel Cell Systems
SOFC	Solid Oxide Fuel Cell
UDP	User Datagram Protocol
WHEx	Water Heat Exchanger

Variables

COP	Coefficient Of Performance
h	heat transfer convective coefficient [W/m^2K]
I	current density [A/m^2]
i, j	indexes for Fig. 18
K_p	surge margin
m	mass flow rate [kg/s]
N	rotational speed [rpm]
n	number of moles
p	pressure [Pa]
P	power [W]
q	heat flux [W]
S	surface [m^2]
T	temperature [K]
TIT	Turbine Inlet Temperature [K]
TOT	Turbine Outlet Temperature [K]
U_f	fuel utilization factor

Greek symbols

β	compressor pressure ratio (total to static)
Δ	variation
ε	recuperator effectiveness
λ	conductivity [W/mK]

Subscripts

0, 1, 2, 3, 4, 5, i	subscripts for Fig. 18
amb	ambient
c	compressor
d	on design
in	inlet
M	measured
s.l.	surge line
t	total

3. The commercial machine

The basic machine is a Turbec T100 PHS Series 3 (Turbec, 2002). It is equipped to operate in stand-alone configuration or connected to the electrical grid. This commercial unit consists of a power generation module (100 kW at nominal conditions), a heat exchanger located downstream of the recuperator outlet (hot side) for co-generative applications, and two battery packages for the start-up phase in stand-alone configuration. Even if the machine is located indoors, the outdoor roof was placed over the machine to use its air pre-filters. Figure 1 shows the plant layout diagram of the micro gas turbine as furnished by the manufacturer in the PHS configuration.

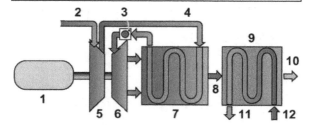

Legend: 1. Generator; 2. Inlet air; 3. Combustion chamber; 4. Air to recuperator; 5. Compressor; 6. Turbine; 7. Recuperator; 8. Exhaust gases; 9. Exhaust gas heat exchanger; 10. Exhaust gas

Fig. 1. Turbec T100 machine: PHS standard layout (courtesy of Turbec).

The power module (Fig. 2) is composed of a single shaft radial machine (compressor, turbine, synchronous generator) operating at a nominal rotational speed of 70000 rpm and a TIT of 950°C (1223.15 K), a natural gas fed combustor, a primary-surface recuperator, a power electronic unit (rectifier, converter, filters and breakers), an automatic control system interfaced with the machine control panel, and the auxiliaries. In this test rig the micro gas turbine is operated using its commercial control system. It works at constant rotational speed when the machine is in stand-alone mode. In this mode the control system changes the fuel mass flow rate to maintain the shaft (in steady-state condition) at 67550 rpm. In grid-connected mode this controller works at constant turbine outlet temperature (TOT). So, in this second mode the control system changes the fuel mass flow rate to maintain the machine TOT (in steady-state condition) at 645°C (918.15 K). However, in both modes the power electronic system allows to generate a 50 Hz current output (at each load values).

Fig. 2. The T100 power module installed at the laboratory of the University of Genoa (power electronics and control system are not shown).

The heat exchanger for water heating (for cogenerative and trigenerative applications) includes an exhaust gas bypass device to control water temperature values. Each start-up battery package includes thirty 12 Volt batteries connected in series for a nominal voltage of 360 Volt DC.

The machine, in its commercial configuration, is equipped with the following essential probes for control reasons: electrical power (±1%), rotational speed (±10 rpm), TOT (±1.5 K), fractional opening values (pilot and main fuel valves), intake temperature, and heating water temperature meters. Furthermore, the commercial machine is equipped with diagnostic probes (e.g. vibration sensor, filter differential pressure meter, temperature probes for the auxiliary systems).

The TPG laboratory was also equipped with a 100 kW resistor bank (pure resistive load) for turbine operation in stand-alone mode. The bank is cooled through an air fan and continuously controlled by an inverter.

4. Machine modifications and connection pipes

The commercial power unit was modified for coupling with the external connection pipes used for flow measurement and management purposes. These modifications are essential for measuring all the properties necessary for cycle characterization (e.g. air mass flow rate or recuperator boundary temperatures), not available in the commercial layout of the machine. For this reason the following modifications were carried out:

- the two pipes between the recuperator and the combustor inlet were substituted with four pipes for the external connections (Fig. 3);
- at the compressor outlet a check valve (Fig. 3) was introduced to prevent damage and block compressor backflow if surge conditions occur during experimental tests;
- between this valve and the recuperator inlet a T-joint was introduced to have a recuperator bypass line which is necessary for studying non-recuperated cycles or hybrid system start-up and shutdown phases (Ferrari et al., 2010a).

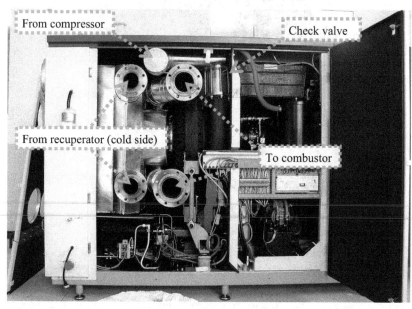

From compressor

Check valve

From recuperator (cold side)

To combustor

Fig. 3. T100 power module modified for coupling with the external pipe system.

The thermally insulated connection pipes are designed for high fidelity flow measurement, achieved with mass flow, pressure, and temperature probes, and flow management carried out with controlled valves. Moreover, these pipes were designed for easy connection to additional components between the compressor and expander, such as the modular vessel presented in the following paragraph. For this reasons, they merge the two flows at the recuperator outlets and split the combustor inlet flow for the connection to the two combustor pipes (see Figs. 3 and 4). Since this kind of mass flow measurements is carried out through pitot devices, each probe is preceded by a pipe of at least 18 diameter length and followed by a minimum length pipe of 4 diameters. This layout was chosen to obtain flow uniformity essential for high-precision measurements. A cold bypass was included to bypass the recuperator or connect the compressor outlet directly to additional external components (e.g. the modular vessel (Ferrari, 2009) used to emulate additional component dimensions). Moreover, the test rig includes a bleed line (Ferrari, 2009), equipped with a globe valve (VB), located at the compressor outlet. This device is essential to bleed part of flow when operative conditions are too close to the surge region. An additional pipe, equipped with a mass flow rate meter was also installed to directly connect recuperator outlet to combustor inlet (through the VM valve), as in a typical recuperated cycle. These pipes are also essential in changing recuperator mass flow rate values.

The test rig was equipped with additional probes to measure the flow properties (Pascenti et al., 2007) (see Table 1). All additional transducers are connected to a PC, through a FieldPoint™ device, and their signals are acquired via LAN using a software developed in LabVIEW™ environment. These measurement devices are used to measure mass flow rates, pressures and temperatures inside the different lines of the plant. The locations of these probes are shown in the test rig layout diagram (Fig. 4). This plant layout refers to the machine working in the standard recuperated cycle (equipped with a check valve, a recuperator bypass, a bleed line, and additional probes).

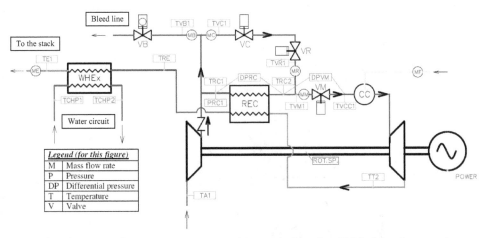

Fig. 4. Plant layout and instrumentation: machine and pipes (see Table 1 for the complete legend).

Since this machine is not designed for connection to additional components between the compressor and expander the connection pipes were developed to incur the lowest possible pressure drops. This is necessary to prevent surge conditions, and to avoid strong machine performance decrease. Therefore, gate valves were chosen (except in the case of the bleed line), and the pipes were designed with wide theoretical support.

A preliminary analysis was carried out with a test rig model in order to evaluate machine performance decrease (due to additional pressure and temperature losses), and to evaluate the operative limits. Then, a CFD analysis (see (Ferrari et al., 2007) for details) was performed with Fluent tool to verify the pipe design.

The test rig model was implemented with TRANSEO tool (in MATLAB®-Simulink® environment). This software (Traverso, 2005) is a simulation program based on an easy access library designed for the off-design, transient and dynamic analyses of advanced energy systems based on microturbine technology. TRANSEO was developed and validated at the University of Genoa in previous studies carried out by both Ph.D. students and TPG researchers (Caratozzolo et al., 2010; Ferrari et al., 2007; Traverso et al., 2005). After an experimental validation in design conditions (Ferrari et al., 2009a), the model was used to calculate the operative curves (at constant TOT) on the compressor map (Fig. 5) at different pressure drop values (Δp) between recuperator and combustor. Each curve was calculated with a valve operating at fixed fractional opening. For this reason, Fig. 5 curves are obtained maintaining constant the ratio between the pressure drop (Δp) and the compressor outlet pressure (p_c). A maximum drop of 485 mbar was calculated at $\Delta p/p_c = 0.108$ to prevent any risk of surge (with the following surge margin limitation: $K_p > 1.1$). To have a wide operative range, as required during the tests, a pressure drop limit of 300 mbar was calculated (the curve at $\Delta p/p_c = 0.069$ has a Δp = 290 mbar at 70000 rpm). These results were necessary to design pipes and valves with wide operative ranges during the tests and manage the rig under safe conditions. The nominal diameter values chosen for these pipes are: (i) 125 mm for the pipes immediately upstream of the combustor, (ii) 100 mm for the other connection pipes. The experimental tests showed good performance in agreement with the design target to obtain the lowest possible additional pressure drop (with the lowest space occupation too). The maximum pressure loss value (DPVM probe) measured during the tests is about 60 mbar.

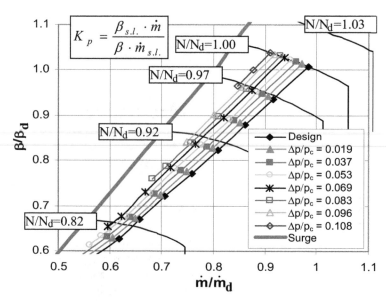

Fig. 5. Machine operation curves at constant Δp/p_c: TRANSEO model calculations.

4.1 Test example: Machine equipped with the external connection pipes

This paragraph shows an example of possible tests carried out on the machine equipped with the external pipes. For this reason Fig. 6 reports (on manufacturer compressor map) the experimental measurements carried out at different load values in stand-alone mode. These tests were performed using the compressor outlet pressure probe (PRC1 to measure the static wall pressure) and a mass flow meter (MM) located in the direct connection between the recuperator outlet and the combustor inlet (Fig. 4).

It is important to highlight these measurements, because they are essential for a correct theoretical analysis of the machine behaviour at both component and system levels and for a complete model validation. While manufacturer compressor maps are usually essential for commercial machine models, this test rig allows to carry out wide experimental verification activities on these performance curves. Moreover, the experimental compressor maps are essential to prevent surge events during tests operated with external additional components (e.g. the modular vessel).

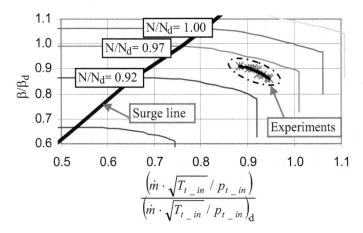

Fig. 6. Direct-line test: experimental data on compressor map obtained in stand-alone mode.

Since in stand-alone conditions the machine control system operates at constant rotational speed, it is possible to measure part of the curve at N/N_d=96.5. Within the accuracy of the probes utilized (Pascenti et al., 2007), the experimental points (Fig. 6) are well located between the 0.97 and 0.92 curves. The difference between the curve obtained from the manufacturer datum interpolation and the experimental points is due to the ambient temperature, which is around 25°C (298.15 K) instead of 15°C (288.15 K), and to the cited probe accuracy. However, ongoing works will be carried out to measure more map points, by converting the data into ISO conditions or operating at the ISO ambient temperature.

5. Machine connected to an external vessel

The machine operation with an additional volume located between compressor outlet and combustor inlet is typical of different innovative plant layout cycles (hybrid system, mHAT, externally fired cycle). This kind of volume is extremely significant on machine transient behaviour especially for plant component safety reasons (to avoid stress or dangerous conditions (Pascenti et al., 2007)). Therefore, this test rig was equipped with a modular vessel designed for experimental tests on the coupling of the machine with different kinds of components (e.g. saturators, fuel cells of different layouts or technology, or additional heat exchangers). The effect of these innovative components can be emulated through a right number of vessel modules to couple the real volume dimension with the micro gas turbine. The emulation of thermal and flow composition aspects can be carried out through the management of test rig valves (Ferrari et al., 2010a), the hardware/software coupling based on real-time models (Bagnasco, 2011), or the injection of additional flows (e.g. for the chemical composition emulation (Ferrari et al., 2011)).

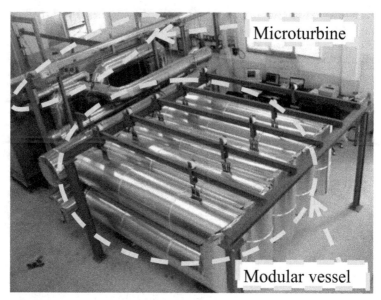

Fig. 7. Modular vessel coupled with the machine.

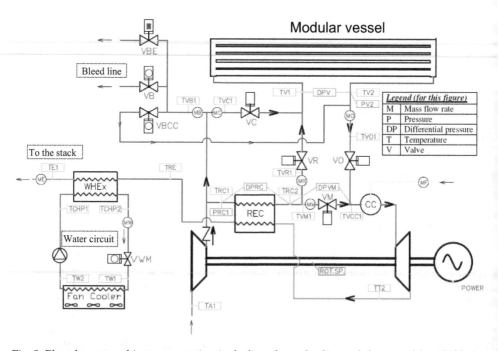

Fig. 8. Plant layout and instrumentation including the cathodic modular vessel (see Table 1 for the complete legend).

Mass flow rates			
Name	Location	Probe type	Accuracy
MM	Main line	Pitot tube	±1%
ME	Plant outlet	Thermal meter	±3%
MF	Fuel inlet	Thermal meter	±1%
MR	Vessel inlet from the recuperator	Pitot tube	±1%
MO	Vessel outlet	Pitot tube	±1%
MC	Vessel inlet from the compressor	Pitot tube	±1%
MB	Bleed outlet	Pitot tube	±1%
MW	Water main line	Magnetic meter	±4%
Static pressures			
Name	Location	Probe type	Accuracy
PA1	Ambient	Ambient sensor	±1%
PRC1	Recuperator inlet	Absolute	±1%
DPRC	Recuperator loss	Differential	±1%
DPVM	Main line loss	Differential	±1%
DPV	Vessel loss	Differential	±1%
PV2	Vessel outlet	Absolute	±1%
Temperatures			
Name	Location	Probe type	Accuracy
TA1	Ambient	Ambient sensor	±1%
TRC1	Recuperator inlet	Thermocouple	±2.5 K
TRC2	Recuperator outlet	Thermocouple	±2.5 K
TVM1	Main line pitot	Thermocouple	±2.5 K
TVCC1	Combustor inlet	Thermocouple	±2.5 K
TT2	Turbine outlet	Thermocouple	±2.5 K
TE1	Plant outlet	Thermocouple	±2.5 K
TVR1	From the recuperator pitot	Thermocouple	±2.5 K
TV1	Vessel inlet	Thermocouple	±2.5 K
TV2	Vessel outlet	Thermocouple	±2.5 K
TVO1	From the vessel pitot	Thermocouple	±2.5 K
TVC1	From the compressor pitot	Thermocouple	±2.5 K
TVB1	Bleed valve inlet	Thermocouple	±2.5 K
TCHP1	WHEx inlet	PT100 RTD	±0.3 K
TCHP2	WHEx outlet	PT100 RTD	±0.3 K
TC1	Compressor inlet	PT100 RTD	±0.3 K
TRE	Recuperator outlet	Thermocouple	±2.5 K
TW1	Cooler inlet	PT100 RTD	±0.3 K
TW2	Cooler outlet	PT100 RTD	±0.3 K

Table 1. Additional probes referred to Fig. 10 layout.

The vessel is composed of two collector pipes, connected to the recuperator outlet and the combustor inlet respectively (see Figs. 7 and 8), and five module pipes connected to both collectors. Both collectors and module pipes have a nominal diameter of 350 millimetres and their total length is around 43 meters for a maximum volume of about 4 m³. This vessel can emulate the volume of additional components suitable for the machine size.

Figure 8 shows the rig layout with the modular vessel and all the additional probes and valves introduced in this plant configuration (see Table 1 for further instrumentation details). In comparison with the layout shown in Fig. 4, a vessel outlet line equipped with a gate valve (VO) was installed, a bleed emergency valve (VBE) was included to prevent surges during emergency shutdown, and an additional globe pneumatic valve (VBCC) was introduced to connect the compressor outlet directly to the combustor inlet. Moreover, the test rig was further improved with the installation of a water fan cooler located outside of the laboratory. It is based on three 0.7 kW electrical fans used to cool down the water (coming from the WHEx of Fig. 8), and to operate in closed circuit conditions (a 1.5 kW variable speed pump was installed).

6. Hybrid system emulation devices

To perform tests related on high temperature fuel cell hybrid systems a part of the external vessel (collectors and four modules) is used to emulate the cathodic dimension of a fuel cell stack. For this reason, the maximum cathodic volume is about 3.2 m³. The fifth vessel module is used to emulate the related anodic volume (about 0.8 m³) and an ejector (Ferrari et al., 2006) based anodic recirculation was included in the rig. Moreover, this facility was equipped with a steam injection system to emulate the turbine inlet composition typical of a hybrid system and a real-time model was connected to the plant for components not physically present in the rig. All these emulation devices were designed to analyse a SOFC based hybrid system of 450 kW electrical load (consistent with the machine size) (Ferrari et al., 2009a) scaled on the basis of the Rolls-Royce Fuel Cell Systems (RRFCS) planar stack (size: 250 kW; fuel utilization: 75%; stack temperatures: 800-970°C; current density: 2940 A/m²) (Massardo & Magistri, 2003). The layout of the test rig equipped with the anodic recirculation and the steam injection systems is shown in Fig. 9.

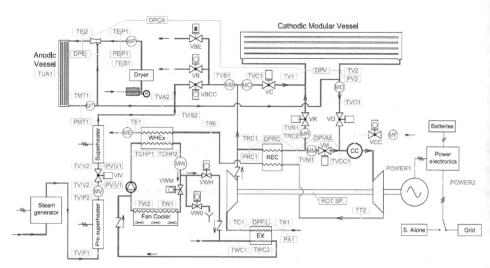

Fig. 9. Plant layout and instrumentation including the anodic recirculation and the steam injection systems.

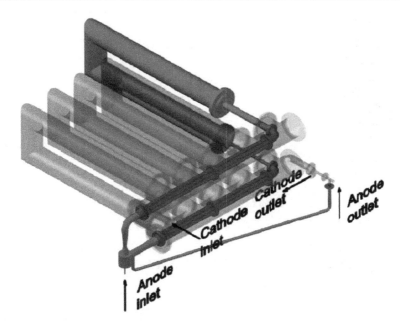

Fig. 10. The anodic loop layout.

6.1 The anodic recirculation and the steam injection systems

The anodic recirculation system (Fig. 10) is composed of a compressed air line (for fuel flow emulation), an anodic single stage ejector, and an anodic vessel. The compressed air line was designed to supply the ejector primary duct with an air mass flow rate up to 20 g/s (Ferrari et al., 2009b). This approach was developed to emulate the fuel mass flow rate at the ejector primary duct with an air mass flow rate. The anodic ejector generates the recirculation flow rate through this system as in a typical SOFC hybrid system. The 0.8 m³ anodic vessel was designed to heat up the flow in the anodic loop. For this reason, to better emulate the anodic side, a pipe based heat exchanger was developed as shown in (Ferrari et al., 2010b). To better show this layout, a 3-D CAD plot (Fig. 10) was developed: part of the anodic loop was inserted into the cathodic vessel to partially heat the anodic flow.

The steam injection system was designed to obtain at the expander inlet the same c_p value of the reference hybrid system. This similitude approach is completely explained in (Ferrari et al., 2011). For this reason the rig was equipped with a 120 kW electrical steam generator capable of at least 27 g/s, a 40 kW electrical super-heater to increase the steam temperature from the steam generator outlet condition to a temperature suitable for the turbine combustor inlet (around 515°C), and a controlled valve for the mass flow rate management. The measuring of the steam flow rate requires a completely mono-phase steam. So upstream of the mass flow probe, an additional electrical heater (called Pre-superheater) was installed. Furthermore, several thermocouples were added for control and diagnostic purposes. The final layout is shown in Fig. 9.

All the additional probes included in the rig for these additional hardware devices (anodic recirculation and steam injection systems) are reported in Tab. 2.

Mass flow rates			
Name	Location	Probe type	Accuracy
MP	Ejector primary line	Thermal meter	±1%
MT	Anodic volume line	Venturimeter	±3%
MV	Steam generator outlet	Vortex meter	±1%
Static pressures			
Name	Location	Probe type	Accuracy
PEjP1	Ejector primary duct inlet	Absolute	±1%
PMT1	Anodic volume line	Absolute	±1%
DPEj	Anodic ejector	Differential	±1%
DPCA	Cathodic/Anodic circuit pressure difference	Differential	±1%
PVIV1	Steam control valve inlet	Absolute	±1%
PVIV2	Steam control valve outlet	Absolute	±1%
Temperatures			
Name	Location	Probe type	Accuracy
TEjP1	Ejector primary duct inlet	Thermocouple	±2.5 K
TEjS1	Ejector secondary duct inlet	Thermocouple	±2.5 K
TEjS2	Ejector outlet	Thermocouple	±2.5 K
TUA1	Anodic volume (U pipe inlet)	Thermocouple	±2.5 K
TMT1	Anodic flow venturimeter	Thermocouple	±2.5 K
TVA2	Anodic circuit outlet	Thermocouple	±2.5 K
TVIP1	Steam generator outlet	Thermocouple	±2.5 K
TVIP2	Pre-superheater generator outlet	Thermocouple	±2.5 K
TVIV1	Steam control valve inlet	Thermocouple	±2.5 K
TVIV2	Steam control valve outlet	Thermocouple	±2.5 K
TVIS2	Steam system outlet	Thermocouple	±2.5 K

Table 2. Additional probes (not shown in Fig. 8 layout) referred to Fig. 9.

6.2 Fuel cell stack real-time model to be connected to the rig

To complete the emulation of a SOFC hybrid system a real-time model was developed in Matlab®-Simulink® to be coupled to the experimental test rig. This model (developed with components validated in previous works (Ghigliazza et al., 2009a; Ghigliazza et al., 2009b) was based on the simplification of simulation components (cell stack, reformer, anodic loop, off-gas burner, expander) developed in the TRANSEO tool (Traverso, 2005) of the TPG research group. Through the Real-Time Windows Target tool and the UDP interface approach it is possible to study the entire hybrid system using the model for components not physically present in the test rig.

Figure 11 shows how the real-time model and the experimental facility are connected to emulate the entire hybrid system. The real-time model receives (as inputs) the mass flow rate, the pressure and the temperature values at the machine combustor inlet level. Furthermore, it receives the machine rotational speed and the acquisition time values. These input data are transferred from LabVIEW™ software to the real-time model (in Simulink®) through an apt UDP interface. The real-time model is used to calculate the fuel cell system behaviour over time. In detail, this model is composed of a reformer, a SOFC stack, an off-gas burner, an anodic ejector based recirculation system, a cathodic blower based recirculation, and the expander of the T100 machine. So, in this work this real-time model is used to calculate the TOT values coming from an interaction between the SOFC system and the T100 expander. This TOT value is used over time to control the real machine (the electrical load in stand-alone mode) to produce the same TOT value of the model. Also this output interface is managed through the UDP approach. Moreover, the interface includes a port to transfer to LabVIEW™ software the mass flow rate values of anodic ejector primary duct (MP). This values are used to carried out tests with the MP mass flow rate equal to the fuel flow calculated in the model. With this hardware/software interconnection layout it is possible to emulate the stack/turbine interaction from an experimental point of view studying different operative conditions.

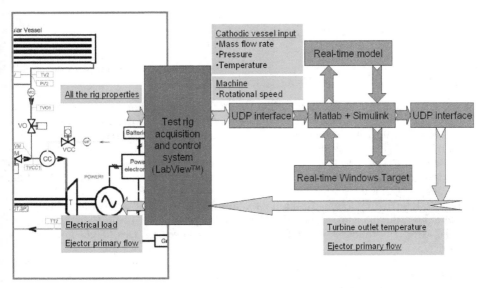

Fig. 11. Hybrid system emulation: hardware/software interconnection layout.

6.2.1 Test example: Hybrid system emulation
This paragraph reports an example of possible tests to be carried out with these emulator devices. This test was performed according to the following procedure:
1. establish connection between the real-time model and the plant;
2. impose requested current and fuel variation on model user interface;
3. the model simulates in real-time mode the evolution of SOFC properties towards a new operative point;
4. the calculated values of TOT and anodic ejector primary flow are continuously fed to the plant control system as set point values;
5. the control system moves machine load until the measured TOT value is equal to the calculated one, and operates on the MP control valve to generate flow values coming from the model.

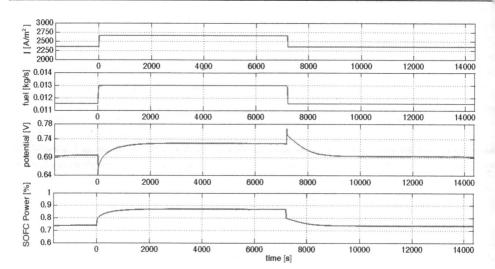

Fig. 12. Emulation test with the real-time model: current density (I), fuel mass flow rate, cell potential, SOFC power (referred to its design value).

The test is based on a concurrent variation of SOFC drawn current and mass flow rate of anodic ejector fuel. From a steady-state condition, corresponding to 80% of nominal current and primary fuel flow, the model performed a step increase in both parameters to 90% of nominal values. When a new steady-state condition (about 2 hours) was reached the model brought back the system to the initial condition. Figure 12 shows the trend of steps related to the input parameters. Moreover, the behaviour of single cell potential, and SOFC total electric power (AC) are reported. Both properties show an initial step variation, due to the input steps, followed by a long duration variation due to the SOFC temperature behaviour over time (high thermal capacitance of the SOFC). In detail, the cell voltage increases when the SOFC temperature increases for the electrical losses decrease (see (Bagnasco, 2011) for further details).

Figure 13 shows the trend of measured turbine outlet temperature and the calculated turbine inlet temperature values: it can be easily noticed how the TOT signal is affected by a constant noise characteristic of values coming from field while the TIT (calculated) shows a smoother behaviour. These values increase with the current and fuel increase (SOFC temperature increase). Figure 13 also shows the variation of the utilization factor (Eq. 1) and hybrid system (HS) global efficiency (Eq. 2). The U_f peak is due to the fluid dynamic delay between the current variation (instantaneous) and the fuel change on the stack. The hybrid system efficiency shows the step effect followed by a long time variation due to thermal aspects. The global efficiency oscillation, with high peaks in the area after the second step (7500-8000 s), are due to unexpected fuel flow discontinuities during the test.

$$U_f = \frac{\left[nH_2 \right]_{cons}}{\left[4 \cdot nCH_4 + 6 \cdot nC_2H_6 \right]_{in}} \tag{1}$$

$$\eta_{HS} = \frac{P_{GT} + P_{FC} - P_{blower}}{\dot{m}_{fuel}} \tag{2}$$

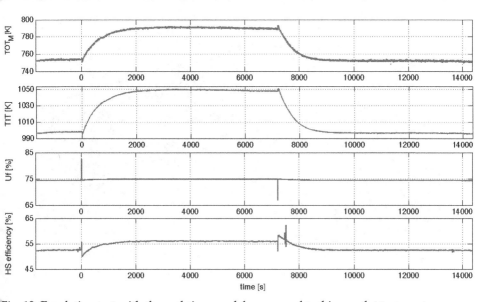

Fig. 13. Emulation test with the real-time model: measured turbine outlet temperature (TOT$_M$), turbine inlet temperature (TIT), fuel utilization factor, hybrid system efficiency.

7. Compressor inlet temperature control devices

A new water system was designed and installed to control the machine compressor inlet temperature (Fig. 9 layout). This system is composed of three air/water heat exchangers installed at the machine air intakes (Fig. 14) and connected to the water system. Even if the water pipe layout was modified for the connection with an absorption cooler (see the following paragraph), this part of the paper describes the previous rig layout that was used for a wide experimental campaign carried out on the machine recuperator. In this past configuration, the water system was equipped with three controlled electrical valves (VWM, VWH, and VWO of Fig. 9). It was possible to cool down the compressor inlet air by means of cold water from the supply system (opening VWO), and to heat up this air flow (closing VWM, and opening VWH) by means of the hot water coming from the machine co-generation system (WHEx). A new control system (see (Ferrari et al., 2010c) for details) was developed to manage these valves for the required temperature generation. With this past layout, maximum cooling performance depended on the supply water temperature (about 22°C in summer, that is 295.15 K), while the only restriction for heating performance is the maximum compressor inlet air temperature for the machine cooling system, that is 40°C (313.15 K). However, in the following paragraph, it is possible to notice that an absorber cooler connected to the system allows to study tri-generation options and use the produced cold flow for higher compressor inlet temperature cooling.

7.1 Test example: compressor inlet temperature control

As an example of possible tests to be carried out with the compressor inlet temperature control devices, this paragraph shows the experimental data measured on the recuperator of this test rig when operating in the machine standard layout. Since the large influence on the

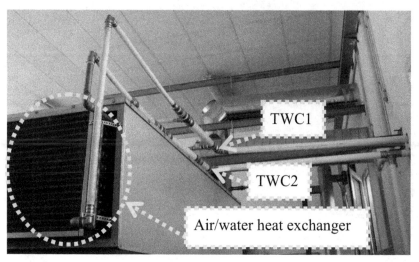

Fig. 14. An air/water heat exchanger with water pipes for compressor inlet temperature control.

recuperator temperatures of the compressor inlet temperature, this new control system was used to maintain this temperature at a fixed value of 28°C (301.15 K, with maximum errors of ±0.3 K during all the tests). These steady-state tests were carried out with the machine connected to the electrical grid to measure recuperator performance at different mass flow rate values. In this configuration the machine control system operates at constant TOT (called TT2 in Fig. 9, that is maintained at 645°C (918.15 K)) and changes the rotational speed (and the air mass flow rate) with load changes. For surge prevention purposes, it is not possible to perform tests below a 20 kW electrical load. After the machine heating phase (during the start-up), the controller does not accept load values below 20 kW.

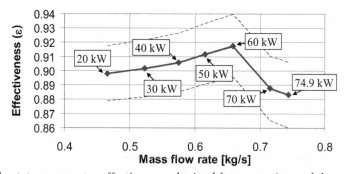

Fig. 15. Steady-state recuperator effectiveness obtained from experimental data at different loads (machine connected to the electrical grid).

Figure 15 shows the recuperator effectiveness (defined in Eq. 3 with Fig. 9 nomenclature) obtained at different electrical load values, i.e. different air mass flow rates. While the continuous line connects the effectiveness values calculated through recuperator boundary temperatures (measured during the tests), the dotted lines show the accuracy values (around ±2%) of this performance parameter. This uncertainty band was calculated through temperature measurements affected by a ±2.5 K accuracy. As shown in previous theoretical works (e.g. (McDonald, 2003)), starting from maximum flow it is possible to observe an effectiveness increase (from 0.883 to 0.918) with the mass flow rate decrease, and an effectiveness maximum followed by a decrease. However, this final trend is not so relevant as in (McDonald, 2003), because machine control system does not enable to operate steady-state tests under 20 kW (under 0.47 kg/s). Further details on all the recuperator temperatures and other tests carried out on this heat exchanger, when operating with the T100 modified machine, are shown in (Ferrari et al., 2010c).

$$\varepsilon = \frac{TRC2 - TRC1}{TT2 - TRC1} \tag{3}$$

8. Test rig integration with an absorber cooler unit

To carry out tests at compressor inlet temperature values under 20°C and to study tri-generative configurations, the facility was equipped with an absorber cooler. This device exploits the thermal content of machine exhaust flow (water at 95°C, that means 368.15 K, produced by the WHEx of Fig. 9) to produce cold water (7-12°C, that means 280.15-285.15 K). This refrigeration energy is essential for the machine inlet and for the laboratory cooling during summer or long time tests.

As shown in Fig. 16 the machine produces, in full load conditions, about 113 kW of thermal power to obtain about 2 l/s of hot water at 95°C (the system operates in closed circuit configuration). With this thermal power the absorber is able to generate about 80 kW of cooling power. This system is based on an absorber inverse cycle (water/lithium bromide) operating (in the test rig conditions) at a 0.7 COP value.

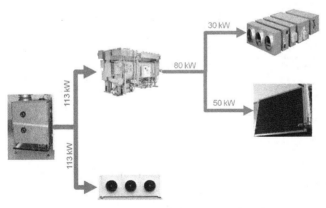

Fig. 16. Maximum thermal power related to both heating and cooling.

The water plant (modified in comparison with Fig. 9) related to refrigeration was designed for a maximum 50 kW cold power for the compressor inlet cooling and 30 kW (at maximum) for the laboratory conditioning. Moreover, with the "Fan Cooler" water/air heat exchanger (already shown in Fig. 9) it is possible to emulate a heating system using a part of the 113 kW thermal power, operating in tri-generative condition. So, While the "Fan Cooler" is used to emulate a heating system, the heat exchangers used to manage the cold power are essential to emulate a cold thermal load.

Figure 17 shows the plant scheme related to cold water generation (from the absorber unit) and thermal power management. For this new water plant three pumps were installed: 1.5 kW pump for the hot water (2 l/s mass flow rate, 2.45 bar pressure increase), 7.5 kW pump for the cold water (5 l/s mass flow rate, 8.34 bar pressure increase), and a third pump (5.5 kW) to refrigerate the condenser of the absorber unit (12 l/s mass flow rate, 2.74 bar pressure increase). A 260 kW evaporative tower was installed for the refrigeration water (see (Prando et al., 2010) for further details). To operate the laboratory cooling a 20 kW water/air heat exchanger was designed and installed in the rig. Moreover, to perform heating conditions at the compressor inlet level, as already included in the previous facility configuration, (for instance for the emulation of a summer performance during winter) a water/water heat exchanger was included. It is a plate exchanger (power: 80 kW, primary flow: 1.44 l/s, secondary flow: 5 l/s) used to heat the "Ex" water directly with the hot water from the "WHEx" (see Fig. 17 for layout details). The water plant was also equipped with controlled valves for flow management and with mass flow rate (Magnetic meter - accuracy: ±4%) and temperature (PT100 RTD - accuracy: ±0.3 K) probes for measurement of main properties (see (Prando et al., 2010) for further details).

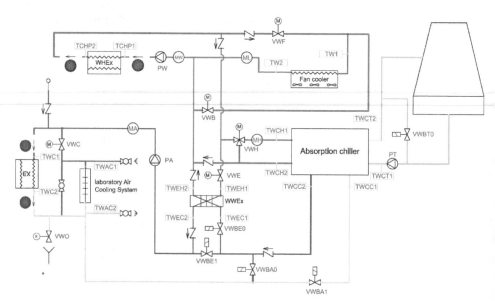

Fig. 17. Water system plant layout for the tests with absorber unit.

9. Model validation activities

Great attention is devoted to this activity to validate time-dependent simulation models at both component (recuperator) and system (the hybrid system emulator test rig) levels. A good level of consistency can be achieved thanks to the complete knowledge of the test rig dimensions, volumes, masses, shaft inertia, thermal capacitances, and operating procedure. Such completeness is difficult to obtain in industrial plants, where details about equipment are often missing or confidential. The following paragraph shows an example of this kind of validation activities focusing the attention on the machine recuperator.

9.1 Test example: The recuperator model

This validation activity regards the primary-surface (cube geometry) recuperator located inside the power case of the Turbec T100 machine (Turbec, 2002). So, a recuperator real-time model was tested against experimental data not in a heat exchanger test rig, but in a real operative configuration, working in a commercial recuperated 100 kWe machine. The recuperator model adopts the lumped-volume approach (Ferrari et al., 2005) for both hot and cold flows. Since momentum equation generates negligible contribution during long-time transients, because it produces quite fast effects (dynamic effects) that are negligible in a component with average flow velocities at around 10 m/s, it is possible to properly represent the transient behaviour of the heat exchanger just using the unsteady form of the energy equation (the actual governing equation (Ghigliazza et al., 2009a) of the system).

The finite difference mathematical scheme (shown in Fig. 18) is based on a recuperator division into four main parts (j = 0, 1, 2, 3). The internal grid is "staggered" to model the heat exchange between each solid cell (j = 1, 3) and the average temperature of the flow (j = 0, 2): M+1 faces correspond to M cells. The resulting quasi-2-D approach is considered a good compromise between accuracy of results and calculation effort. The heat loss to environment and the longitudinal conductivity into solid parts are also included. All the equations and the integration approach of this model are described in (Ghigliazza et al., 2009a). Moreover, this paper reports the main data used for the recuperator model (Table 3) for the results reported here.

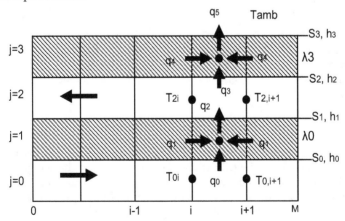

Fig. 18. Real-time model: finite difference scheme.

Figure 19 shows the comparison between experimental data and model results related to recuperator outlet temperature (cold side). The test considered here is a machine start-up phase carried out from cold condition. The results obtained during this test are acceptable, even if same margin of improvement exists. With reference to Fig. 19, the following aspects can be highlighted:

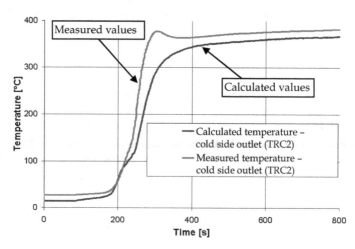

Fig. 19. Recuperator model validation (start-up phase): cold side outlet.

Thermal capacitance	226.05 [kJ/K]
Convective heat exchange (cold side)	500 [W/m²K]
Convective heat exchange (hot side)	250 [W/m²K]
Length	0.35 [m]
Nominal pressure drop (cold side)	0.06 [bar]
Nominal pressure drop (hot side)	0.06 [bar]

Table 3. Recuperator model data.

- the matching between measured data and model predictions is within a difference of 50°C, which can be considered a good result considering real-time simulation performance;
- measurements show a longer thermal delay (likely explanation: effect due to thermal shield of thermocouples).

10. Conclusion

A new test rig based on micro gas turbine technology was developed at the TPG laboratory (campus located at Savona) of the University of Genoa, Italy. It is based on the coupling of different equipments to study advanced cycles from experimental point of view and to provide students with a wide access to energy system technology. Particular attention is devoted on tests related on hybrid systems based on high temperature fuel cells. The main experimental facilities developed and built for both student and researcher activities are:

- A commercial recuperated micro gas turbine (100 kW nominal electrical load) equipped with a hot water co-generation unit and with the essential instrumentation for control reasons and to operate typical tests (start-up, shutdown, load changes) on the machine.
- A set of external pipes connected to the machine for the flow measurement and management. These pipes are used to measure with enough accuracy all the properties necessary for cycle characterization (e.g. the air mass flow rate or recuperator boundary

temperatures), not available in the machine commercial layout. In particular the chapter shows a test example related to the compressor map measuring.

- An external modular vessel to test the coupling of the machine with different additional innovative cycle components, such as saturators, fuel cells of different layouts or technology, or additional heat exchangers.
- Additional devices for hybrid system emulation activities. This part describes the anodic recirculation based on a single stage ejector (coupled to the rig for tests related to the anodic/cathodic side interaction), the steam injection system based on a 120 kW steam generator (used to emulate the turbine inlet composition typical of a hybrid system), and a real-time model used to emulate the components not physically present in the rig (e.g. the fuel cell). As an example of tests carried out with these devices, this chapter reports the main results obtained during fuel and current steps carried out with the real-time model coupled with the rig.
- Compressor inlet temperature control devices (heat exchangers, pipes, pump, and control system) to evaluate performance variations related to ambient temperature changes. Particular attention is focused on tests carried out on the recuperator with the machine operating in grid-connected conditions.
- An absorber unit connected to the plant (the hot water generated by the WHEx is used as primary energy to produce cold water) to carry out tests at compressor inlet temperature values under 20°C and to study tri-generative configurations.
- Great attention is devoted to validation activities for time-dependent simulation models. As an example, this chapter shows the comparison between experimental data and model results related to recuperator outlet temperature (cold side), during a cold start-up phase.

Besides the additional developments and tests on the rig, already planned and presented in (Pascenti et al, 2007; Ferrari et al., 2009a; Ferrari et al., 2010c; Prando et al., 2010), all the different layout configurations will be considered for tests. For instance, in an ongoing work it is planned to use the real-time model for control system development activities related on SOFC hybrid plants and the absorber cooler to carry out tests at lower ambient temperature conditions, also considering tri-generative configurations.

11. Acknowledgment

This test rig was mainly funded by FELICITAS European Integrated contract (TIP4-CT-2005-516270), coordinated by Fraunhofer Institute, by LARGE-SOFC European Integrated Project (No. 019739), coordinated by VTT, and by a FISR National contract, coordinated by Prof. Aristide F. Massardo of the University of Genoa.

The authors would like to thank Prof. Aristide F. Massardo (TPG Coordinator) for his essential scientific support, Dr. Loredana Magistri, (permanent researcher at TPG) for her activities in design point definition, and Mr. Alberto N. Traverso (associate researcher at TPG) for his technological support on absorption cooler installation.

12. References

Kolanowski, B. F. (2004). *Guide to Microturbines*, Fairmont Press, ISBN 0824740017, Lilburn, Georgia (USA).

Boyce, M. P. (2010). Handbook for Cogeneration and Combined Cycle Power Plants, Second Edition, ASME Press, ISBN 9780791859537, New York, New York (USA).

Massardo, A. F., McDonald, C. F., & Korakianitis, T. (2002). Microturbine-Fuel Cell Coupling for High-Efficiency Electrical-Power Generation. Journal of Engineering for Gas Turbines and Power, Vol. 124(1), pp. 110-116, ISSN 0742-4795.

Lindquist, T., Thern, M., Torisson, T. (2002). Experimental and Theoretical Results of a Humidification Tower in an Evaporative Gas Turbine Cycle Power Plant. *Proceedings of ASME Turbo Expo 2002*, 2002-GT-30127, ISBN 0791836010, Amsterdam, The Netherlands, June 3-6, 2002.

Traverso, A., Massardo, A. F., Scarpellini, R. (2006). Externally Fired micro-Gas Turbine: Modelling and Experimental Performance. *Applied Thermal Engineering*, Elsevier Science, Vol. 26, pp. 1935-1941, ISSN 1359-4311.

Magistri, L., Costamagna, P., Massardo, A. F., Rodgers, C., McDonald, C. F. (2002). A Hybrid System Based on a Personal Turbine (5 kW) and a Solid Oxide Fuel Cell Stack: A Flexible and High Efficiency Energy Concept for the Distributed Power Market, *Journal of Engineering for Gas Turbines and Power*, Vol. 124, pp. 850-875, ISSN: 0742-4795, New York, New York (USA).

Magistri, L., Traverso, A., Cerutti, F., Bozzolo, M., Costamagna, P., Massardo, A. F. (2005). Modelling of Pressurised Hybrid Systems Based on Integrated Planar Solid Oxide Fuel Cell (IP-SOFC) Technology. *Fuel Cells*, Topical Issue "Modelling of Fuel Cell Systems", WILEY-VCH, Vol. 1, Issue 5, ISSN 1615-6854.

Pedemonte, A. A., Traverso, A., Massardo, A. F. (2007). Experimental Analysis of Pressurised Humidification Tower For Humid Air Gas Turbine Cycles. Part A: Experimental Campaign. *Applied Thermal Engineering*, Elsevier Science, Vol. 28, pp. 1711–1725, ISSN 1359-4311.

McDonald, C. F. (2003). Recuperator Considerations For Future High Efficiency Microturbines. *Applied Thermal Energy*, Elsevier Science, Vol. 23, pp. 1453-1487, ISSN 1359-4311.

Ferrari, M. L. (2011). Solid Oxide Fuel Cell Hybrid System: Control Strategy for Stand-Alone Configurations. *Journal of Power Sources*, Elsevier, Vol. 196, Issue 5, pp. 2682-2690, ISSN: 0378-7753.

Tucker, D., Liese, E., Gemmen, R. (2009). Determination of the Operating Envelope for a Direct Fired Fuel Cell Turbine Hybrid Using Hardware Based Simulation. *Proceedings of International Colloquium on Environmentally Preferred Advanced Power Generation 2009*, ICEPAG2009-1021, ISBN 3-7667-1662-X, Newport Beach, California, USA.

Hohloch, M., Widenhorn, A., Lebküchner, D., Panne, T., Aigner, M. (2008). Micro Gas Turbine Test Rig for Hybrid Power Plant Application. *Proceedings of ASME Turbo Expo 2008*, GT2008-50443, ISBN 0791838242, Berlin, Germany.

Ferrari, M. L., Pascenti, M., Bertone, R., Magistri, L. (2009a). Hybrid Simulation Facility Based on Commercial 100 kWe Micro Gas Turbine. *Journal of Fuel Cell Science and Technology*, Vol. 6, pp. 031008_1-8, ISSN: 1550-624X, New York, New York (USA).

Ferrari, M. L., Pascenti, M., Magistri, L., Massardo, A. F. (2010a). Hybrid System Test Rig: Start-up and Shutdown Physical Emulation, *Journal of Fuel Cell Science and Technology*, Vol. 7, pp. 021005_1-7, ISSN: 1550-624X, New York, New York (USA).

Ferrari, M. L., Pascenti, M., Magistri, L., Massardo, A. F. (2010b). Analysis of the Interaction Between Cathode and Anode Sides With a Hybrid System Emulator Test Rig, *Proceedings of International Colloquium on Environmentally Preferred Advanced Power Generation 2010*, ICEPAG2010-3435, Costa Mesa, CA (USA).

Turbec T100 Series 3 (2002). Installation Handbook.

Pascenti, M., Ferrari, M. L., Magistri, L., Massardo, A. F. (2007). Micro Gas Turbine Based Test Rig for Hybrid System Emulation, *Proceedings of ASME Turbo Expo 2007*, GT2007-27075, ISBN: 0791837963, Montreal, Canada.

Traverso, A. (2005). TRANSEO Code for the Dynamic Performance Simulation of Micro Gas Turbine Cycles, *Proceedings of ASME Turbo Expo 2005*, GT2005-68101, ISBN: 0791846997, Reno, Nevada (USA).

Traverso, A., Calzolari, F., Massardo, A. F. (2005). Transient Behavior of and Control System for Micro Gas Turbine Advanced Cycles, *Journal of Engineering for Gas Turbine and Power*, Vol. 127, pp. 340-347, ISSN: 0742-4795, New York, New York (USA).

Ferrari, M. L., Liese, E., Tucker, D., Lawson, L., Traverso, A., Massardo, A. F. (2007). Transient Modeling of the NETL Hybrid Fuel Cell/Gas Turbine Facility and Experimental Validation, *Journal of Engineering for Gas Turbines and Power*, Vol. 129, pp. 1012-1019, ISSN: 0742-4795, New York, New York (USA).

Caratozzolo, F., Traverso, A., Massardo, A. F. (2010).Development and Experimental Validation of a Modelling Tool for Humid Air Turbine Saturators, *Proceedings of ASME Turbo Expo 2010*, ASME Paper GT2010-23338, ISBN: 9780791838723, Glasgow, UK.

Bagnasco, M. (2011). Emulation of SOFC Hybrid System With Experimental Test Rig and Real-Time Model, Bachelor Thesis, TPG, Genova, Italy (in Italian).

Ferrari, M. L., Pascenti, M., Traverso, A. N., Massardo, A. F. (2011). Hybrid System Test Rig: Chemical Composition Emulation With Steam Injection, *Proceedings of International Conference on Applied Energy*, pp. 2821-2832, Perugia, Italy.

Ferrari, M. L., Bernardi, D., Massardo, A. F. (2006). Design and Testing of Ejectors for High Temperature Fuel Cell Hybrid Systems, *Journal of Fuel Cell Science and Technology*, Vol. 3, pp. 284-291, ISSN: 1550-624X, New York, New York (USA).

Massardo, A. F., Magistri, L. (2003). Internal Reforming Solid Oxide Fuel Cell Gas Turbine Combined Cycles (IRSOFC-GT) – Part II: Energy and Thermoeconomic Analyses, *Journal of Engineering for Gas Turbines and Power*, Vol. 125, pp. 67-74, ISSN: 0742-4795, New York, New York (USA).

Ferrari, M. L., Pascenti, M., Magistri, L., Massardo, A. F., (2009b). Hybrid System Emulator Enhancement: Anodic Circuit Design, *Proceedings of International Colloquium on Environmentally Preferred Advanced Power Generation 2009*, ICEPAG2009-1041, Newport Beach, California, USA.

Ghigliazza, F., Traverso, A., Pascenti, M., Massardo, A. F. (2009a). Micro Gas Turbine Real-Time Modeling: Test Rig Verification", *Proceedings of ASME Turbo Expo 2009*, GT2009-59124, Orlando, Florida (USA).

Ghigliazza, F., Traverso, A., Massardo, A. F., Wingate, J., Ferrari, M. L. (2009b). Generic Real-Time Modeling of Solid Oxide Fuel Cell Hybrid Systems, *Journal of Fuel Cell Science and Technology*, Vol. 6, pp. 021312_1-7, ISSN: 1550-624X, New York, New York (USA).

Ferrari, M. L., Pascenti, M., Magistri, L., Massardo, A. F. (2010c), Micro Gas Turbine Recuperator: Steady-State and Transient Experimental Investigation, *Journal of Engineering for Gas Turbines and Power*, Vol. 132, pp. 022301_1-8, ISSN: 0742-4795, New York, New York (USA).

Prando, A., Poloni, M. (2010). Absorber Tri-generative System Based on a 100 kWe Micro Gas Turbine: Study and Plant Development, Bachelor Thesis, TPG, Genova, Italy (in Italian).

Ferrari, M. L., Traverso, A., Magistri, L., Massardo, A. F. (2005). Influence of the Anodic Recirculation Transient Behaviour on the SOFC Hybrid System Performance, *Journal of Power Sources*, Elsevier, Vol. 149, pp. 22-32, ISSN: 0378-7753.

7

Afterburning Installation Integration into a Cogeneration Power Plant with Gas Turbine by Numerical and Experimental Analysis

Ene Barbu et al*

National Research and Development Institute for Gas Turbines COMOTI,
Romania

1. Introduction

Gas turbines applications represent a continuous challenge for engineers regarding the design, manufacturing and efficient working at nominal and partial loads with respect to environmental and future development requirements. The development of gas turbines in terms of performances as well as of lifetime and safe running has made these installations preferable for cogenerative processes and the lower maintenance costs have lead to particularly tempting recovery terms, down to 3-4 years for a 15-20 years lifetime. The increase in the overall efficiency of the cogenerative group is affected by the degree of use of the heat produced by the gas turbine along with the flue gases. The recovery of the flue gases heat is achieved by using a heat recovery steam generator, usually for producing steam (or hot water). The performances of a heat recovery steam generator depend on the gas turbine's working regime which makes the steam parameters difficult to control. The trend in the field of heat recovery steam generator and afterburning installations are related to the development in gas turbines. The increase in the temperatures of the gas turbine requires new materials that withstand operating regimes in terms of appropriate pollutantsnorms. The oxygen concentration in the gas turbine's flue gases is usually of 11 – 16 % volume. The fact that the combustion process in the gas turbine consumes only a small part of the oxygen from the intake air flow makes possible the application of a supplementary firing (afterburning) for increasing the steam flow rate of the heat recovery steam generator. In aviation the afterburning is used for increasing the thrust of supersonic aircrafts equipped with gas turbines. Introducing the afterburning in cogenerative applications leads to increasing the flexibility and the overall efficiency of the cogenerative group. The burner of the afterburning installation is usually placed between the gas turbine and the heat recovery steam generator, immersed in the flue gases exhausted by the gas turbine, resulting in a complex gas turbine – afterburning – heat recovery steam generator system. This placement results in the afterburning installation affected by the gas turbine and affects the working of the heat recovery steam generator. Usually, when the heat recovery steam generator is used for delivering superheated steam, the working conditions of the afterburning installation

* Valeriu Vilag, Jeni Popescu, Silviu Ionescu, Adina Ionescu, Romulus Petcu, Cleopatra Cuciumita, Mihaiella Cretu, Constantin Vilcu and Tudor Prisecaru
National Research and Development Institute for Gas Turbines Comoti, Romania

influence mainly the superheater of the heat recovery steam generator (with effects on temperature, flow rate etc.). The design of the gas turbine – afterburning installation – heat recovery steam generator system must take into account these variables for insuring the steam parameters required by the technological process. The active control of the combustion is a concept already accepted and the new generation of afterburning installations will need to answer to the requirements of the new "smart" aggregates which automatically take into account the emissions, the energetic efficiency and the process requirements (PIER, 2002). For that purpose the researches conducted at Suplacu de Barcau 2xST 18 Cogenerative Power Plant has focused on the afterburning installation as integral part of the cogenerative group in terms of stack emissions, superficial temperature profile and power quality (Barbu et al., 2010) as well as on its interaction with the heat recovery steam generator.

2. General principles of the mathematical modelling of the thermo-gas-dynamic and chemical processes in the combustion chambers

The classical approach of the combustion chambers study assumes as a general rule the embracing of a steady character of the phenomena taking place in these installations, constituting only a quasi-adequate manner to the problem of analysing the unsteady phenomena generating important collateral effects. The physical-chemical phenomena succeeding in the combustion chamber are extremely complex, each of them (injection, atomization, vaporization, diffusion, combustion) rigorously depending on the physical factors such as air excess, gases pressure, temperature and velocity in the chamber. It may be admitted that the combustion is normal as long as the fluctuations detected in the combustion chamber only depend on the local conditions and they are randomly distributed in the chamber. The high level of complexity of the phenomena, associated to the flow instabilities, the heat transfer and the combustion reactions, makes them inaccurate to model using simplified mathematical models which only globally consider the processes and which are only slightly dependent on the combustion chamber geometry, the combustion configuration, the walls' screening or the intermediary reactions in the flame. Therefore, here is studied the complex and coupled problem of mathematical modelling for pulsating flow (numerical integration of Navier-Stokes equations with a closing model application), the influence on heat transfer (considering the radiation and convection), the combustion reactions (applying complex combustion mechanisms with high number of reactions and intermediary chemical compounds). The generalized model accurately tracking the complex processes in the combustion chamber may be developed as a group of modules, each associated to a phenomenon (flow, heat transfer, combustion reactions and dispersion phase evolution). This modular approach method allows the separate development of several sub-models with higher accuracy for a certain class of problems. Hence the problem of „closing" the equations system describing the studied phenomenon may and has been solved by using several turbulence models: k-ε (standard, realisable or RNG), Reynolds-stress model, LES - large eddy simulation (high scale modelling), or lately, due to the increase in calculation efforts, DNS – direct numerical simulation.

2.1 Mathematical models used for simulating flow, heat transfer and combustion in combustion chambers

There are two fundamentally different manners used for describing the fluid flow equations: the Lagrangian and Eulerian formulations. From the Lagrangian formulation perspective the flow field represents the movement of small, adjacent fluid elements interacting through pressure and viscous forces. The movement of each fluid element is made according to Newton's second law. This method is however impractical because of the high number of mass elements necessary for reaching a reasonable accuracy in describing the flow in a continuous environment. On the other hand the Lagrangian method deserves to be taken into consideration for biphasic flows (gas-droplet type) in describing the dispersion phase because the particles naturally constitute individual mass elements. The complexity of the Eulerian formulation of the biphasic flow does not allow the direct application of the solving schemes existing in the case of monophasic flow. As a consequence of this averaging problem in most numerical models the Lagrangian formulation is used for describing the dispersion phase. The relation between the Lagrangian and the Eulerian formulations is given by the Reynolds transport theorem. Therefore the components on the three axes may be combined in a single vectorial equation:

$$\rho \frac{D\mathbf{u}}{Dt} = \rho \mathbf{F} - \text{grad } p + \mu \nabla^2 \mathbf{u} + \frac{1}{3}\mu \cdot \text{grad}(\text{div}\mathbf{u}) \tag{1}$$

Equation (1) represents the Navier-Stokes in complete vectorial form describing the movement of a viscous fluid. The Navier-Stokes equation is applied for laminar flows as well as for turbulent flows. However it cannot be directly used in solving the problems associated to turbulent flow because it is impossible to track the minor fluctuations of the velocity associated with the turbulence. In order to determine the flow field, the numerical model solves the mass and impulse conservation equations. For flows involving heat transfer of compressibility additional equations are needed for energy conservation. For flows implying chemical compounds mixing or chemical reactions an equation of chemical compounds conservation is solved or, in the „probability density function" cases (generically called PDF models), conservation equations for the considered mixture fractions as well as equations defining their variations are needed. In flows with turbulent character additional transport equations need to be solved. Mass conservation equation, or continuity equation, may be written:

$$\frac{\partial \rho}{\partial t} + \frac{\partial}{\partial x_i}(\rho u_i) = S_m \tag{2}$$

Equation (2) is the generalized formulation of mass conservation equation and is applicable for incompressible or compressible flows. The source term S_m represents the mass added to the continuous phase, mass resulted from the dispersion phase (due to liquid droplets' vaporization) or a different source. For axi-symmetric bi-dimensional flows, the continuity equation in given by:

$$\frac{\partial \rho}{\partial t} + \frac{\partial}{\partial x}(\rho u) + \frac{\partial}{\partial r}(\rho v) + \frac{\rho v}{r} = S_m \tag{3}$$

where x is the axial coordinate, r is the radial coordinate, u is the axial velocity and v is the radial velocity. The impulse conservation on i in an inertial reference plane is described by:

$$\frac{\partial}{\partial t}(\rho u_i) + \frac{\partial}{\partial x_j}(\rho u_i u_j) = -\frac{\partial p}{\partial x_i} + \frac{\partial \tau_{ij}}{\partial x_j} + \rho g_i + F_i \tag{4}$$

where p is the static pressure, τ_{ij} is the tension tensor and ρg_i and F_i are the internal and external gravitational forces (e.g. occurring from the interaction with the dispersion phase) on direction i. F_i also includes other source terms depending on the model (such as for porous environment case). The tension tensor τ_{ij} is given by:

$$\tau_{ij} = \left[\mu \left(\frac{\partial u_i}{\partial x_j} + \frac{\partial u_j}{\partial x_i} \right) - \frac{2}{3} \mu \frac{\partial u_l}{\partial x_l} \delta_{ij} \right] \tag{5}$$

where μ is the dynamic viscosity and the second term in the right side of the relation represents the effect of volume dilatation. For axi-symmetric bi-dimensional geometries the axial and radial impulse conservation equations are given by:

$$\frac{\partial}{\partial t}(\rho u) + \frac{1}{r}\frac{\partial}{\partial x}(r\rho uu) + \frac{1}{r}\frac{\partial}{\partial x}(r\rho vu) = -\frac{\partial p}{\partial x} +$$
$$+ \frac{1}{r}\frac{\partial}{\partial x}\left[r\mu\left(2\frac{\partial u}{\partial x} - \frac{2}{3}(\nabla \cdot \vec{v}) \right) \right] + \frac{1}{r}\frac{\partial}{\partial r}\left[r\mu\left(\frac{\partial u}{\partial r} + \frac{\partial v}{\partial x} \right) \right] + F_x \tag{6}$$

$$\frac{\partial}{\partial t}(\rho v) + \frac{1}{r}\frac{\partial}{\partial x}(r\rho uv) + \frac{1}{r}\frac{\partial}{\partial x}(r\rho vv) = -\frac{\partial p}{\partial r} + \frac{1}{r}\frac{\partial}{\partial x}\left[r\mu\left(\frac{\partial v}{\partial x} + \frac{\partial u}{\partial r} \right) \right] +$$
$$+ \frac{1}{r}\frac{\partial}{\partial r}\left[r\mu\left(2\frac{\partial v}{\partial r} - \frac{2}{3}(\nabla \cdot \vec{v}) \right) \right] - 2\mu\frac{v}{r^2} + \frac{2}{3}\frac{\mu}{r}(\nabla \cdot \vec{v}) + \rho\frac{w^2}{r} + F_r \tag{7}$$

where $\nabla\vec{v} = \frac{\partial u}{\partial x} + \frac{\partial v}{\partial r} + \frac{v}{r}$ and w is the tangential velocity. The energy conservation equation may be written:

$$\frac{\partial}{\partial t}(\rho E) + \frac{\partial}{\partial x_i}\left[u_i(\rho E + p) \right] = \frac{\partial}{\partial x_i}\left(k_{ef}\frac{\partial T}{\partial x_i} - \sum_{j'} h_{j'}J_{j'} + u_j(\tau_{ij})_{ef} \right) + S_h \tag{8}$$

where k_{ef} is the effective conductivity ($k_{ef} = k + k_t$, where k_t is turbulent thermal conductivity, defined relative to the utilized turbulence model) and $J_{j'}$ is the diffusive flow of the chemical compound j'. The first three members in the left side of equation (8) represent the energy transfer due to conduction, chemical compounds diffusion and respectively viscous dissipation. The term S_h includes the heat exchanged in the chemical reactions or other volume heat sources. In equation (8),

$$E = h - \frac{p}{\rho} + \frac{u_i^2}{2} \tag{9}$$

where the sensible enthalpy h is defined, for ideal gases, by:

$$h = \sum_{j'} m_{j'} h_{j'} \tag{10}$$

and for incompressible flows by:

$$h = \sum_{j'} m_{j'} h_{j'} + \frac{p}{\rho} \tag{11}$$

In equations (10) and (11) $m_{j'}$ is the mass fraction of the chemical compound j' and

$$h_{j'} = \int_{Tref}^{T} c_{p,j'} \cdot dT \tag{12}$$

is the corresponding enthalpy of the compound, and the reference temperature is $T_{ref} = 298.15\ K$. In combustion studies, when the PDF based nonadiabatic model is used, the model requires solving an equation for total enthalpy, set by the energy equation:

$$\frac{\partial}{\partial t}(\rho H) + \frac{\partial}{\partial x_i}[\rho u_i H] = \frac{\partial}{\partial x_i}\left(\frac{k_t}{c_p}\frac{\partial H}{\partial x_i}\right) + \tau_{ik}\frac{\partial u_i}{\partial x_k} + S_h \tag{13}$$

In the hypothesis of a unitary Lewis number ($Le = 1$), the conduction and diffusion terms of the chemical compounds are combined in the first term in the left side of equation (13), while the viscous dissipation contribution in the nonconservative formulation occurs as the second term of the equation. Total enthalpy H is defined by:

$$H = \sum_{j'} m_{j'} H_{j'} \tag{14}$$

where $m_{j'}$ is the mass fraction of the chemical compound j' and

$$H_{j'} = \int_{Tref}^{T} c_{p,j'} \cdot dT + h_{j'}^{0}(T_{ref,j'}) \tag{15}$$

$h_{j'}^{0}(T_{ref,j'})$ is the enthalpy of formation of chemical compound j' at the reference temperature T_{ref}. Equation (8) includes the terms of pressure and kinetic energy work, terms neglected in incompressible flows. The decoupled solving method for the flow equations does not require including these terms in incompressible flows. However these terms must always be considered when using coupled solving method or for compressible flows. Equations (8) and (13) include the viscous dissipation terms representing the thermal energy created by the viscous tension in the flow. When using the decoupled solving method, the energy equation formulation does not need to explicitly include these terms because the viscous heating is in most cases neglected. The viscous heating becomes important when the Brinkman number, B_r, is close or higher than the unitary value, where

$$B_r = \frac{\mu U_e^2}{k\Delta T} \tag{16}$$

and ΔT represents the temperature difference in the system. The compressible flows usually have a Brinkman number $B_r \geq 1$. In the same time equations (8) and (13) include the enthalpy transport effect due to chemical compounds diffusion. For the decupling solving method the term $\frac{\partial}{\partial x_i}\sum_{j'} h_{j'} J_{j'}$ is included in equation (8), and in the nonadiabatic combustion model (PDF) this term does not explicitly appear in the energy equation, being included in the first term in the right side of equation (13). The energy sources S_h include in equation (8) the energy due to chemical reactions.

$$S_{h,reaction} = \sum_{j'} \left[\frac{h_{j'}^0}{M_{j'}} + \int_{T_{ref,j'}}^{T_{ref}} c_{p,j'} \cdot dT \right] \cdot R_{j'} \tag{17}$$

where $h_{j'}^0$ is the enthaply of formation of compound j', and $R_{j'}$ is the volumetric velocity of creation of compound j'. When using the PDF combustion model, the heat of formation is included in the enthalpy definition so the energy sources of the chemical reaction are no longer included in the formulation of S_h.

3. Suplacu de Barcau 2xST 18 cogenerative power plant

Suplacu de Barcau 2xST 18 Cogenerative Plant (fig. 1), with beneficiary SC OMV PETROM SA, is located in Bihor County, Romania, 75 km from Oradea Municipality. The main technical data are given in table 1. The plant was integrally commissioned in 2004 working in the framework of Suplacu de Barcau Oil Field. The electrical energy is used for driving the reducing gear boxes from the oil wells, the compressors, the pumps, for lighting etc. and the thermal energy (steam) is injected in the deposit being necessary in the oil extraction technological process and/or for other field requirements (heating the buildings or technological pipes). Suplacu de Barcau 2xST 18 Cogenerative Plant comprises two groups (fig. 1, right) which may work together or separately. Each group includes a ST 18 gas turbine (fig. 2, left), an afterburning installation (fig. 2, centre), a heat recovery steam generator (fig. 2, right) and additional installations. The heat recovery steam generator of each cogenerative group is a fire tube type boiler with two flue gas lines – one horizontal and the other vertical, comprising: the uncooled afterburning chamber; the superheater insuring the 300 °C steam temperature; the pressure body producing the saturated steam; the feed water heater assembly – water pre-heater insuring the necessary parameters of the water supplying the pressure body. The superheater is a coil type heat exchanger with 12 coil pipes (ø 38) welded in the steam inlet down-tanks (upper tank – fig. 3, centre) of the pressure body and superheated steam outlet (lower tank – fig. 2, right). The steam in the pressure body enters the upper tank through a PN40 DN150 connector (placed in the middle of the tank) and is distributed to the 12 coil pipes, then enters the lower tank and is delivered to the users.

The gases from the afterburning chamber follow the horizontal line of the heat recovery steam generator (superheater – pressure body) then the vertical one (feed water heater – water pre-heater – stack). Each cogenerative group is able to work in any of the three versions given in table 1, but the basic one is version I. 2xST 18 Cogenerative Power Plant is working automatically, the exploitation personnel being alerted by the command panel, through optical signalling and alarm horns, regarding the deviations of the supervised parameters or the damages occurrence. Certain parameters (pressures, temperatures, flow rates etc) of the equipments are archived and displayed with the help of an acquisition system.

Fig. 1. 2xST 18 Cogenerative Power Plant view (left) and gear placement (right)

No	Name	Version I	Version II	Version III
1	**Cogenerative group version**	Gas turbine + afterburning + heat recovery steam generator	Gas turbine + heat recovery steam generator	Heat recovery steam generator + afterburning
2	Fuel	Natural gas		
3	Gas turbine type	ST 18 (Pratt & Whitney - Canada)		
4	Boiler type	Fire tube boiler (SC UTON SA Onesti - Romania)		
5	Electric generator type	GSI-F (Electroputere Craiova - Romania)		
6	Electrical power delivered by the plant	2x1,75 MW (6,3 kV, 50 Hz)		-
7	Superheated steam pressure	20 bar		
8	Superheated steam temperature	300 ^{0}C		

Table 1. Technical specifications of 2xST 18 Cogenerative Power Plant

Fig. 2. ST 18 gas turbine (left), afterburning installation burner (centre) and heat recovery steam generator (right) at 2xST 18 Cogenerative Power Plant

Fig. 3. Superheater of the heat recovery steam generator at 2xST 18 Cogenerative Power Plant

3.1 The afterburning installation at Suplacu de Barcau 2xST 18 cogenerative power plant

The afterburning installation (burner with automatics) at 2xST 18 Cogenerative Power Plant was delivered by Saacke (www.saacke.com – Germany) and has the specifications given in table 2. The burner (fig. 4), produced by Eclipse – Holland, is the "FlueFire" type dedicated to this kind of application. It may be placed directly in the flue gases flow, between the turbine and the recovery boiler, but may work as well on fresh air. The burner has 21 basic modules located on 3 natural gas fuelling ramps and 2 flame propagation modules. The flue gases from the gas turbine are introduced in the "FlueFire" burner through an adaptation section. The mixture with the fuel is obtained through the swirling motion of the flue gases exhausted from the turbine in the fuel jets. This leads to the cooling and the stabilization of the combustion in the burner front allowing downstream high temperatures at a low NO_x content. The air, delivered by a fan (fig. 4, right), is introduced in the adaptation section through a distribution system built to insure an uniform distribution in the transversal section because the emissions depend on the unevenness of the flow, velocity, oxygen concentration etc. The afterburner modules are built in refractory steel, laser cut, for insuring the necessary uniformity.

Each module is fitted in the natural gas fuelling ramps using two gas nozzles in order to allow the free dilatation of the assembly. The ignition is initiated with the help of a pilot burner placed in the lower area of the afterburning burner and the supervision of the flame is insured by a UV type "DURAG D-LX 100 UL" detector placed in the upper area.

No.	Name		Value
1	Natural gas pressure	Before the regulator	0,5-2 bar
		After the regulator	0,4 bar
2	Thermal power	With flue gases (version I)	2,4 MW
		With air at 20 ºC (version III)	6 MW
3	Flue gases maximum mass flow rate		8,75 kg/s
4	Flue gases temperature at the inlet of the burner		524 ºC
5	Flue gases temperature at the end of the afterburning chamber (versions I and III)		770 ºC

Table 2. Technical specifications regarding the afterburning installation at 2xST 18 Cogenerative Power Plant

 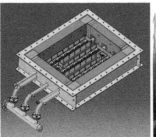 

Fig. 4. The burner in the afterburning installation at 2xST 18 Cogenerative Power Plant (left, centre) and the fresh air fan (right)

4. Gas turbine – afterburning interaction

The flue gases flow in the outlet section of the gas turbine is generally turbulent and unevenly distributed. In some areas at the inlet of the afterburning installation backflows may occur. A uniform flow distribution is an important factor concurring to the good working of the afterburning and to the performances of the heat recovery steam generator. Grid type burners are designed to distribute the heat uniformly in the transversal section of the heat recovery steam generator, fact requiring careful oxygen feeding in order to avoid high NO_x emissions and variable length flame. The flow rate, the temperature, the composition of the flue gases exhausted from the gas turbine depend on the fuel type, load, fluid injection in the gas turbine (water, steam), environmental conditions etc. The gas turbines used in industrial applications are fuelled by liquid or gaseous fuels. Regarding the liquid fuels, for economical reasons, there are usually used cheap fuels such as heavy fuels, oil fuels or residual products from different manufacturing processes or chemisation (Carlanescu et al., 1997). Using these types of fuels raises problems concerning: insuring combustion without coating, decreasing the corrosive action caused by the presence of aggressive compounds (sulphur, traces of calcium, lead, potassium, sodium, vanadium) and problems concerning pumping and spraying (heating, filtration etc.). When considering using aviation gas turbines for industrial purposes (existing aviation gas turbines with minimal modifications) the possibilities of using liquid fuels are limited. For each case the technical request of the beneficiary must be analyzed in conjunction with the study of fuel characteristics affecting the processes in the combustion chamber (density, molecular mass, damping limits, burning point, volatility, viscosity, superficial tension, latent heat of vaporization, thermal conductibility, soot creation tendency etc.). For the gaseous fuels the problem is easier considering the high thermal stability, the absence of soot and ashes, and the high caloric power. In this case the problems concern mostly the combustion process in conjunction with the requirements of the used gas turbine. For the valorisation of the landfill gas the TV2 – 117A gas turbine was modified to work on landfill gas instead of kerosene by redesigning the combustion chamber (Petcu, 2010). Numerical simulations and experimentations have been conducted for the gas turbine working on liquid fuel (kerosene) and gaseous fuels (natural gas, landfill gas). The boundary conditions have been either calculated or delivered by the gas turbine manufacturer for three working regimes: take-off, nominal and idle. The results are presented in table 3 with the corresponding temperatures for each regime. The temperature fields are displayed in rainbow with red representing the highest value. The most important result refers to the fact that the numerically obtained temperatures are close enough to the ones indicated by the manufacturer, the differences being explained by the simplifying hypotheses introduced in the simulations. Analyzing the numerical results it may be observed that the flame shortens (column 5) with the decrease in regime, but it fills better the area between two adjacent injectors (column 6). The main criterion validating the numerical results has been the averaged turbine inlet temperature (T_m) in the conditions of fuel and air flow rate imposed by the working regimes of the TV2 – 117A gas turbine. For working on gaseous fuel, the TV2 – 117A gas turbine has suffered adjustments on the fuel system level and particularly on the injection nozzles. The starting point in designing the new injection nozzle was a previous application on the TA2 gas turbine resulted by modifying the TV2 – 117A. Table 4 presents the variation of the CH_4 mass fraction indicating the injected jet shape (left) and the burned gases temperature in the combustion chamber outlet/turbine inlet section (right) for different working regimes. The geometrical parameters of the injection nozzle were set based on the numerical temperature fields in the turbine inlet section and aiming to obtain a compact fuel jet which avoids the combustion chamber walls. It must be noted that a stable combustion process has been obtained using a gaseous fuel in a combustion chamber designed for a different type of fuel (kerosene). The numerical simulations made possible narrowing the variation domains for the geometrical and gas-dynamic parameters in order to establish the constructive solution of the combustion chamber for working on landfill gas. The numerical results have been used for designing and manufacturing the new injection nozzle for the eight injectors of the TV2 – 117A gas turbine, transforming it into the TA2 aero-derivative. From tables 3 and 4 it may be noticed that by changing the fuel and the working regime the temperature distribution in the section of interest is modified, fact that might affect the afterburning installation performances. For reducing the NO_x emissions, reducing the temperature in the combustion area is applied through water or steam injection.

Regime	T_m [K]	Thermal field at the outlet	Thermal field on the walls	Thermal field in the axial-median section	Thermal field on the frontier between two sections
1	2	3	4	5	6
Take-off	1135				
Nominal	1075				
Cruise	1039				

Table 3. Numerical results for the TV2 – 117A gas turbine on liquid fuel (Petcu, 2010)

No.	T_m [K]	Fuel injection jet	Combustion chamber outlet temperature field
1	1053		
2	952		
3	944		

Table 4. Numerical results for the TA2 gas turbine on landfill gas - injection jet and thermal cross section from numerical simulations (Petcu, 2010)

In some cases the water or steam are injected directly in the combustion area through a number of holes in the combustion chamber inlet section or in the fuel injection nozzle. Another solution is injecting the water upstream of the firing tube, usually in the air flow that next passes the turbulence nozzles in its way to the combustion area. This method insures a very good atomization, the small droplets being transported by the air flow while the larger ones form a thin film on the surface of the turbulence nozzles being next atomized by the air passing over the downstream edges of the turbulence nozzle. The efficiency of the water or steam injection in reducing the NO_x emissions has been highlighted by many authors and may be expressed by the following relation (Carlanescu et al., 1998):

$$\frac{NO_{x(wet)}}{NO_{x(dry)}} = \exp-\left(0.2x^2 + 1.41x\right) \tag{18}$$

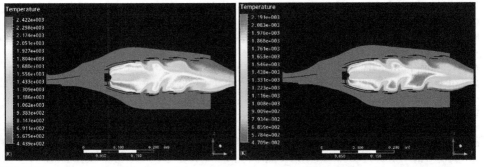

Fig. 5. Temperature field in the axial-median section of the TA2 combustion chamber, T_m =1063 K, without water injection (left) and with water injection (right) (Popescu et al., 2009)

Relation (18) may be applied for liquid as well as for gaseous fuels, showing that approximately 80 % reduction in the NO_x emissions may be obtained at equal water/steam-to-fuel flow rates ($x = 1$). The water injection is more efficient at higher combustion pressures and temperatures where the NO_x production is higher and less efficient at lower pressures and temperatures. For the independent running of the TA2 gas turbine on methane, without afterburning, the numerical simulations (fig. 5 - 6) regarding the water injection have shown a NO_x reduction of over 50 % (Popescu et al., 2009).

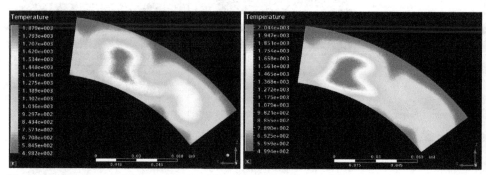

Fig. 6. Temperature field in the combustion chamber outlet section, T_m =1063 K, without water injection (left) and with water injection (right) (Popescu et al., 2009)

However, theoretical and experimental researches on a turbojet have shown that the steam injection reduces the NO emissions up to 16% (mass fractions) when a steam flow which doubles the fuel flow is introduced (Benini et al., 2009). At the same ratio, the NO reduction in the water injection case is approximately 8%. The steam injection slightly reduces the CO level while the water injection raises it with the increase in the injected water quantity. Using the NASA CEA program (McBride & Gordon, 1992; Zehe et al., 2002), the combustion have been analyzed in the TA2 gas turbine for methane, natural gas (the composition at Suplacu de Barcau Cogenerative Plant) and landfill gas (equal volume proportions of methane and carbon dioxide) in the pressure and temperature conditions recommended for different working regimes of the gas turbine (T_m = 1063, 1023 and 873 K). The air excess coefficients have been established for each fuel in the dry working cases in order to obtain the same temperature of the reaction products for a corresponding fuel quantity. Starting from these initial data and increasing the injected water quantity up to 50 % of the fuel quantity, a decrease in temperature has been noticed for each 10 % injected water of approximately 1.46 degrees for methane, 1.62 degrees for natural gas and 1.36 degrees for landfill gas. It was then aimed to establish the dependence of the quantitative water-to-fuel ratio at supplementary fuel injection in order to maintain the maximum temperature in the gas turbine. These analysis have been made only for methane for the same stable working regimes of the TA2 gas turbine (T_m = 1063, 1023 and 873 K). In the theoretical calculations for methane some limitations have been next applied: a ratio between the fuel quantity in the water injection case and the initial fuel quantity of maximum 2; a minimum oxygen concentration in the flue gases from the gas turbine of 11 % volume for afterburning running. The general reaction for methane combustion when water injection is involved is given by equation (19) and the algorythm used in the NASA CEA program for determining the water injection influence is presented in fig. 7:

$$b \cdot CH_4 + 2(O_2 + 3.76N_2) + a \cdot H_2O \rightarrow c \cdot H_2O + d \cdot CO_2 + e \cdot N_2 + f \cdot O_2 + g \cdot CO + h \cdot NO_x \quad (19)$$

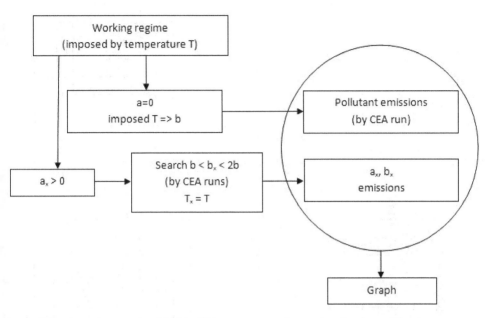

Fig. 7. The algorithm used in NASA CEA program to determine the water injection influence (subscript „x" denominates the seeked coefficients for the desired fixed temperature)

The thermodynamic properties of the system have been tracked with focus on the reaction products concentrations and particularly CO and NO_x. In these conditions the calculations have been made for a water coefficient (denoted a) of maximum 8 (fig. 8). For a higher value than 6.5 an instability of the curves occurs. For a value of the coefficient of approximately 2.8, the gas turbine at the regime $T_m = 1063$ K is close to the minimum oxygen limit of 11% needed for the afterburning. In fig. 9 and 10 the water coefficient a is limited to 4. Therefore we may have a maximum water coefficient of 2.8 for the regime $T_m = 1063$ K, 3.1 for the regime $T_m = 1023$ K and 3.5 for the regime $T_m = 873$ K, having as result the inaccessible areas of emissions reduction for the TA2 gas turbine with water injection when using the afterburning. The theoretical calculations indicate that the NO_x emissions for the TA2 working at the regime $T_m = 1063$ K with afterburning may not be lower than 40 ppm. The unevenness of the flow when exiting the combustion chamber (tables 3 and 4, fig. 5 and 6) and the variation in the burned gases composition (fig. 8 – 10) affect the afterburning process and confirm the necessity of a fine control on the injected water quantity particularly when a significant reduction of the emissions is aimed. Therefore the afterburning is affected from the efficiency, emissions, flame stability points of view as well as from the corrosion on the elements subjected to the burned gases action.

The combined actions of water vapours and oxygen concentration and high temperature of the flue gases represent the recipe for an accentuated corrosion (Conroy, 2003).

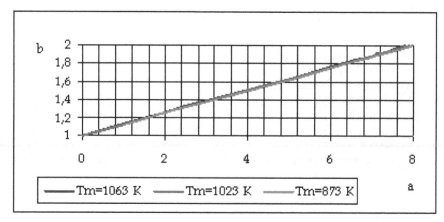

Fig. 8. Fuel coefficient (b) variation depending on injected water coefficient (a)

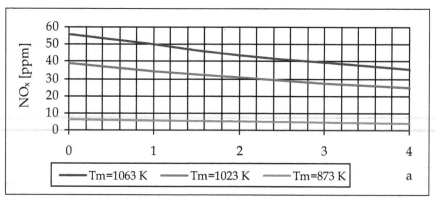

Fig. 9. NO_x concentration variation depending on injected water coefficient (a)

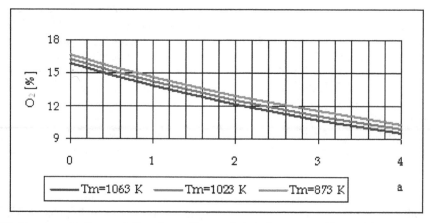

Fig. 10. O_2 concentration variation depending on injected water coefficient (a)

5. Afterburning – Heat recovery steam generator interaction

The heat recovery steam generator in the cogenerative groups is designed to run at certain parameters of the flue gases and superheated steam. The process requirements, the variable environmental conditions (affecting the gas turbine and therefore the afterburning) influence the running of the heat recovery steam generator and the overall efficiency of the cogenerative group. The steam superheater is usually the last heat exchanger in the pressurized thermal circuit water – steam of a heat recovery steam generator. As „end of the line" element it is its duty to maintain the temperature of the superheated steam, imposing mechanical systems and automatics with well defined functions and roles taken into consideration by the designer. Increasing the nominal parameters of modern heat recovery steam generator as a result of the increased performances in gas turbines led to a superheater area larger than that of the vaporizing system, the superheater becoming a large metal consumer as well as the heat exchanger with the highest thermal demand. As consequence, the superheater needs the proper consideration in the design process as well as in activity. If the flow process in the superheater pipes is admitted as isobar, the increase in temperature takes place according to an exponential law. The value of the coefficient of thermal unevenness in the flue gases flow entering the superheater stage depends on the constructive shape and the thermal diagram. This coefficient is defined by (Neaga, 2005):

$$k'_t = \frac{t'_{g\,max} + 273}{t'_{g\,med} + 273} \tag{20}$$

where the index shows that the temperatures are indicated for the inlet of the stage, $t'_{g\,max}$ and $t'_{g\,med}$ being the maximum and respectively the averaged temperatures of the gases at the inlet. The ability of the steam to absorb the heat of the flue gases in different areas of the volume occupied by the stage leads to unevenness affecting the safe running of the heat exchanger. The researches show that the highest unevenness on the outlet of the stage is registered in the counter-stream flow of the thermal agents regardless of the number of stages of the superheater and the lowest in the uniflow (Neaga, 2005). The characteristics of the superheated steam are different for the radiation and convection superheaters. The radiation superheater absorbs more heat at low loads while the convection one absorbs more heat at higher loads (Ganapathy, 2001). The superheaters usually have more stages, the radiation and convection combined ensuring a uniform temperature in the steam for a larger range of loads. When the superheater consists in only one stage the problem of controlling the temperature in the superheated steam becomes more complicated. The steam temperature can usually be maintained constant in the 60 – 100 % load range but several factors act on the superheated steam temperature: heat recovery steam generator load, air excess coefficient at the furnace outlet, initial dampness of the fuel, calorific value of the fuel etc. As consequence, the temperature control systems must comply with certain conditions: low inertia, large control range (regardless of the variable parameter leading to variations in the superheated steam temperature), safe running construction, minimal manufacturing and running expenses etc.

6. Researches concerning the integration of the afterburning installation with the gas turbine and the heat recovery steam generator at Suplacu de Barcau 2xST 18 cogenerative plant

6.1 The integration of the afterburning installation with the heat recovery steam generator

The researches for the integration of the afterburning installation with the heat recovery steam generator and the gas turbine at Suplacu de Barcau 2xST 18 Cogenerative Power Plant have been made in several stages. At commissioning, the functional tests at some working regimes of the heat recovery steam generator have shown high temperatures of the superheated steam (table 5) compared to the nominal temperature (300 °C), leading to frequent activations of the heat recovery steam generator. In these conditions the coil pipes have been counted, from 1 to 12 (fig. 11), in the direction of the steam circulation through the lower tank, the outer temperature has been measured (fig. 3, 11) with a contact thermometer (Barbu et al., 2006) and the exchange area of the superheater has been reduced by replacing the first three coil pipes with three L-shaped pipes in order to maintain the steam velocity. Fig. 12 presents the temperature distribution on the outer surface of the 12 coil pipes of the superheater's outlet tank (t_{eSI}) after the replacement, for the averaged temperature of the superheated steam t_{vSI} = 280 °C and the temperature of the gases at the afterburning chamber outlet t_{ca}=690 °C.

Coil pipe	2	4	6	8	10	12	Observations
Outer temperature [°C]	370	380	295	295	245	185	Before replacement; t_{vSI}= 322 °C
	210	398	366	364	212	198	After replacement; t_{vSI}= 280 °C

Table 5. Outer temperature of the coil pipes of the superheater's outlet tank

The data in table 5 and fig. 12 (left) shows a modification in temperature distribution after the replacement of the three coil pipes. In the end, the pipes 1 – 4 have been cut and plugged and refractory metal sheets have been applied on each superheater (for a better circulation of the flue gases in the superheater) leading to the temperature distribution presented in fig 12 (right). Fig. 12 (right) illustrates an increase in the thermal field evenness in the coil pipes compared to the initial case at the commissioning of the heat recovery steam generator (table 5). For the process requirements of superheated steam temperature up to 350 °C only the area of the screens has been adequately reduced.

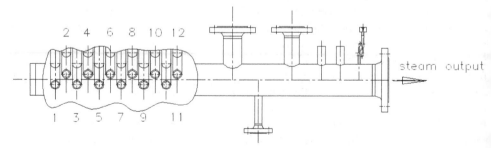

Fig. 11. Counting of the coil pipes of the superheater's outlet tank

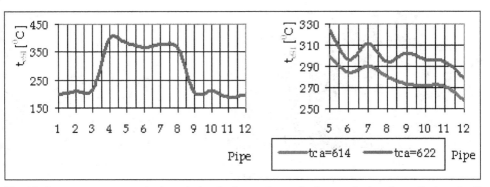

Fig. 12. Outer temperature (t_{eSI}) variation in the outlet tank after replacing the coil pipes 1 – 3 with three L-shaped pipes (t_{vSI}= 280 ^{0}C, t_{ca}=690 ^{0}C - left) and after plugging the coil pipes 1 – 4 and installing the screens on the superheater (t_{ca}=614 ^{0}C s and 622 ^{0}C - right)

6.2 The integration of the afterburning installation with the gas turbine
The data obtained until present resulted from experimentations conducted at 2xST 18 Cogenerative Plant, numerical simulations in CFD environment and experimentations with natural gas and air on the test bench.

6.2.1 The afterburning integrated analysis system
After solving the problem of the superheated steam temperature control, the next step has been simulating in CFD environment the afterburning installation at Suplacu de Barcau 2xST 18 Cogenerative Plant (Barbu et al., 2010). The numerical results have indicated that the air (or flue gases from the gas turbine) distribution in the burner may be improved leading to the installation of a concentrator at group 1. The second cogenerative group has remained unmodified since the commissioning. For determining the influence of the concentrator on the combustion process several aspects have been analysed: emissions at the stack, noise, superficial temperature profile and power quality for different working regimes of the groups. The measurements have been performed in industrial conditions on both cogenerative groups before applying overall optimization solutions in order to not disturb the technological process. For this reason the measurements have been performed for partial loads. For noise analysis have been used three measuring chains and a software application for acoustic prediction according to 2002/49/EC Directive, offering an image of the noise propagation in the area of interest. The noise measurements at the afterburning installation have been performed with a 01dB Metravib SOLO sound level meter. The acoustic field in the station's area has been studied in 50 measuring points with the B&K 2250 sound level meter and the acoustic pressure level of the cogenerative group has been determined with the multi-channel acquisition system 01dB Metravib EX-IF10D/module IEPE with 12 microphones 40AE G.R.A.S.. The measurements concerning the quality of environmental air have been performed with the help of the mobile laboratory (fig. 13, left) especially equipped for the task. For the chemical measurements has been used a Horiba PG 250 gas analyzer with the probe installed at the stack (fig. 13, centre). The outer superficial temperature profile has been determined with the help of a Fluke infrared camera, Ti45FT type, the sighting being in the upper area of the burner (fig. 13, right).

  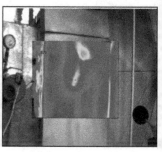

Fig. 13. Mobile laboratory at INCDT COMOTI (left), emissions measurements at the stack (centre) and sighting area of Fluke camera (Ti45FT type) with sound level meter (right)

The process parameters of the gas turbine, heat recovery steam generator and afterburning installation have been displayed in the command room or locally. The correlation of the emissions, noise, superficial temperature profile and power quality has been made related to the time of measurements. The electro-energetic measurements have been made in the electric generator cell through measurement converters, the equipment consisting in devices fixed in panels (ammeters, voltmeters, active and reactive energy counters) and mobile devices (CA 8332B analyser for electro-energetic network and power quality).

6.2.2 Experimental data and numerical simulation of the afterburning at 2xST 18 cogenerative plant

For starters a noise map for cogenerative group 2 has been established (without air concentrator), with group 1 out of work (fig. 14), in order to acquire comparison data for the case of local recording of noise at the burner of the afterburning installation. In the burner area the noise level has been in the 80 – 85 dB range with a significant distortion of noise curves and high values in the area of turbo-generators room. The measurements on group 1 (with concentrator) regarding emissions, noise and outer superficial temperature profile have been performed for the case of the heat recovery steam generator running with the afterburning on fresh air.

Fig. 14. Suplacu de Barcau 2xST 18 Cogenerative Plant – noise map for group 2, with group 1 out of work

Emissions, noise and outer superficial temperature profile measurements have been performed for group 1 (with concentrator) for the case of the heat recovery steam generator running with the afterburning on fresh air, with group 2 working in cogeneration (gas turbine + afterburning + heat recovery steam generator). The set of experimental data has been obtained for five working regimes defined by the gases temperature at the outlet of the afterburning chamber (t_{ca}): 500 ^{0}C, 552 ^{0}C, 604 ^{0}C, 645 ^{0}C, 700 ^{0}C. The NO_x variation and the noise locally measured at the burner (fig. 13) for group 1 (heat recovery steam generator + afterburning on fresh air) depending on flue gases temperature are given in fig. 15. The flue gases temperature has been measured with a thermocouple at the outlet of the afterburning chamber.

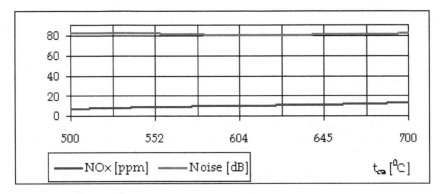

Fig. 15. NO_x and noise variation depending on flue gases temperature – group 1 (heat recovery steam generator + afterburning on fresh air)

Fig. 15 illustrates an increase in NO_x emissions with the flue gases temperature, while the noise level is approximately constant at 80 dB. For the outer superficial temperature profile, according to fig. 13 (right), an increase may be noticed in the area of the isotherms above 200 °C (fig. 16, left) with the increase in the temperature at the outlet of the afterburning chamber from 500 °C to 645 °C. The measurements on group 2 (without concentrator) regarding emissions, noise, outer superficial temperature and power quality have been performed for two cases: without afterburning (gas turbine + heat recovery steam generator) and with afterburning (gas turbine + afterburning + heat recovery steam generator). Another two sets of experimental data have been obtained. One set corresponds to three working regimes of the gas turbine + heat recovery steam generator version, defined by the flue gases temperature at the outlet of the afterburning chamber (t_{ca}): 423 ^{0}C, 437 ^{0}C, 475 ^{0}C. Another set corresponding to the version gas turbine + afterburning + heat recovery steam generator led to the following temperatures (t_{ca}): 536 ^{0}C, 569 ^{0}C, 605 ^{0}C, 645 ^{0}C. The configuration of the isotherms for the gas turbine + afterburning + heat recovery steam generator version is given in fig 16 (right). Working at fresh air rating, more than 645 ^{0}C, the region occupied by the isotherms is reduced. This area is however larger than the one for group 2 (without concentrator – fig 16, right) in gas turbine + afterburning + heat recovery steam generator version or gas turbine + heat recovery steam generator version. The central areas occupied by the isotherms (near the afterburning chamber) for group 2 (without concentrator) are decreasing with the increase in flue gases temperature at the outlet of the afterburning chamber. As opposite, for group 1, as stated above, the areas occupied by the isotherms are increasing with the increase in flue gases temperature.

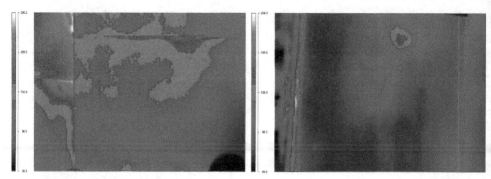

Fig. 16. Isotherms configuration for groups 1 and 2 in infrared; t_{ca} = 645 ⁰C; left – group 1 (with concentrator): heat recovery steam generator + afterburning on fresh air; right – group 2 (without concentrator): gas turbine + afterburning + heat recovery steam generator

Local noise measurements near the burner confirm the values in fig. 14. The figures 17 – 19 present the electro-energetic measurements, respectively the power variations at the generator's hubs and the distortion coefficients from the fundamental (THD) for current and voltage depending on the flue gases temperature at the outlet of the afterburning chamber.

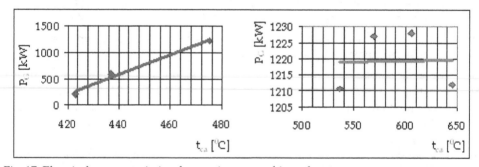

Fig. 17. Electrical power variation for version gas turbine + heat recovery steam generator (left) and version gas turbine + afterburning + heat recovery steam generator (right) depending on the flue gases temperature at the outlet of the afterburning chamber

In version gas turbine + heat recovery steam generator the variation in the heat recovery steam generator load is achieved through the variation of the gas turbine parameters. The electrical power increases with the increase in flue gases temperature (fig. 17, left). For the version gas turbine + afterburning + heat recovery steam generator the heat recovery steam generator load is varied through the afterburning parameters which makes the power quasi-constant (approximately 1220 kW) while the temperature at the outlet of the afterburning chamber increases (fig. 17, right). For the version gas turbine + heat recovery steam generator the value of the distortion coefficients from the fundamental (THD) for current and voltage decreases with the increase in temperature at the outlet of the afterburning chamber (fig. 18, 19 left).

This decrease is more pronounced for the currents (fig. 18, left) indicating an aggravation in the power quality. For version gas turbine + afterburning + heat recovery steam generator the value of the distortion coefficients from the fundamental for current and voltage decreases only slightly (under 1%) with the increase in the flue gases temperature at the outlet of the afterburning chamber (fig. 18, 19 right).

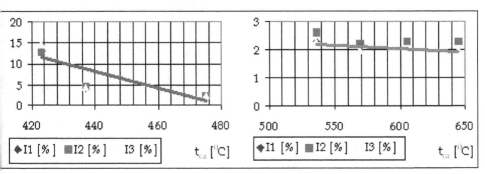

Fig. 18. Variation of distortion coefficients from the fundamental (THD) for current depending on the flue gases temperature at the outlet of the afterburning chamber for version gas turbine + heat recovery steam generator (left) and version gas turbine + afterburning + heat recovery steam generator (right)

For version gas turbine + heat recovery steam generator the variation of the distortion coefficients from the fundamental for current is higher than 5 % (fig. 18, left). This occurs at temperatures at the outlet of the afterburning chamber below 470 °C. The experiments conducted at 2xST 18 Cogenerative Plant have shown an improvement of the flow in the burner section at group 1, particularly in the upper area, but have not allowed an assessment of the performances due to geometric modification of the ST 18 burning module. Based on the experimental data obtained at Suplacu de Barcau 2xST 18 Cogenerative Plant, a new geometry has been obtained for the burning module (ST 18-R, fig. 20 left) analysed through numerical simulations in CFD environment. The new module incorporates an air (flue gases) concentrator and the module ST 18 has been angled to 15° (module ST 18-15) leading to an increased turbulence and a better mixing of the gas fuel with the comburent (Barbu et al., 2010).

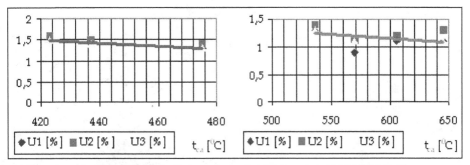

Fig. 19. Variation of distortion coefficients from the fundamental (THD) for voltage depending on the flue gases temperature at the outlet of the afterburning chamber for version gas turbine + heat recovery steam generator (left) and version gas turbine + afterburning + heat recovery steam generator (right)

The numerical simulations (fig. 20, right) indicate NO_x emissions three times lower for the ST 18-R module compared to the old ST 18 model at nominal working regime temperature of the afterburning at 2xST 18 Cogenerative Power Plant (770 ⁰C).

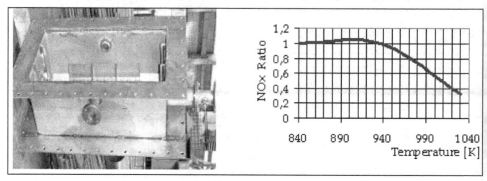

Fig. 20. Burner with three burning modules ST 18-R (left) and the variation of the NO_x ratio depending on the flue gases temperature for modules ST 18 and ST 18-R (right)

6.2.3 Experimental data obtained on test bench

For a thorough investigation of the processes and for eliminating some disturbing factors from the plant test bench examinations were required for the data obtained at 2xST 18 Plant as well as for the numerical results obtained in CFD environment. For this purpose there was designed and manufactured the gas fuel burner INCDT APC 1MGN – UPB (fig. 21, left) with a thermal power of approximately 350 kW, allowing the testing of only one ST 18 or ST 18-15 burning module and adaptable for other geometrical configurations in the same overall dimensions. The natural gas is introduced through a connector in the lower area and the air through a lateral flanged connector. The experiments for the ST 18 (or angled ST 18-15) module took place on the test bench of University Politehnica Bucharest (UPB), Department of Classic and Thermo-mechanical Nuclear Equipment (fig. 21, right). The tests have been made with natural gas and fresh air for different natural gas flow rates (0.5; 0.7 and 0.88 m³/h). The experimental cell was a rectangular enclosure (890x890x990 mm) with a truncated pyramid segment connected to the exhaust stack. The walls are made of glass allowing the observation of the flame, one side door providing access (fig. 21, centre and right).

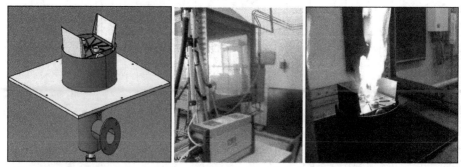

Fig. 21. Burner INCDT APC 1MGN – UPB with module ST 18-15, 3D design (left) and UPB test bench testing (centre and right)

The connection of the burner to the natural gas network was made through a hose and the one to the air fan through a removable flanged assembly. The placement of the burner was achieved through a plate fixed by screws. The emission measurements have been performed with a gas analyser MRU - Analyzer Vario Plus Ind. (fig. 21 centre) and the noise has been monitored with a sound level meter 01 dB Metravib SOLO mounted in the upper area of the enclosure. The outer superficial temperature profile has been determined with the help of a type Ti45FT Fluke infrared camera, the flame being sighted with the access door open (fig. 21, left). The measurements for emissions and noise at the three flow rates have been performed with the access door open. The temperature distribution in the flame has been determined with the help of three thermocouples (type PtRh30% - PtRh46%) placed on a holder on the height estimated for the flame development. The counting of the thermocouples starting from the base was: T_{FL1} , T_{FL2} and T_{FL3}. It was noticed a more homogenous distribution and a slower increase in temperature in the flame on ST 18-15 module, as seen in fig. 22. This also results from table 6 presenting parameters in the infrared recording – module ST 18 (left) and ST 18-15 (right). The areas occupied by the high temperature isotherms are significantly increased for module ST 18-15. Compared to module ST 18, the flame fills more the burning point, it shortens, the temperature distribution is more homogenous and the NO_x and CO emissions decrease. For the flow rate of natural gas of 0.88 m^3/h a decrease in the NO_x emissions occurs, over 30 % for module ST 18-15 compared to ST 18 (fig. 23). However the values of noise for the module ST 18-15 are higher than for ST 18 due to increased turbulence (fig. 24). This phenomenon may be better observed particularly for high flow rates (0.88 m^3/h). The researches conducted until present have shown the superiority of module ST 18-15, the numerical results being validated by the experimental ones obtained on the test bench.

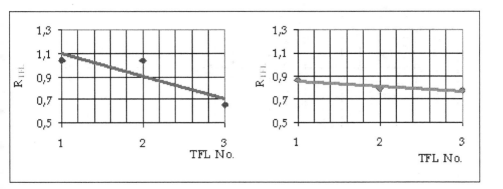

Fig. 22. Variation of the temperature ratios in the flame (for T_{FL1} , T_{FL2} and T_{FL3}), corresponding to modules ST 18-15 and ST 18 at Q_{combs} = 0.5 m^3/h (left) and Q_{combs} = 0.88 m^3/h (right)

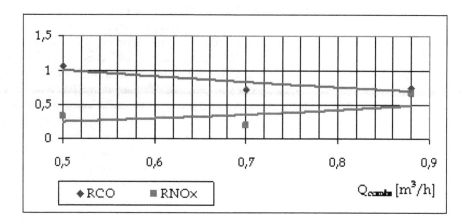

Fig. 23. The CO and NO$_x$ ratios variation corresponding to ST 18-15 and ST 18 modules depending on natural gas volume flow

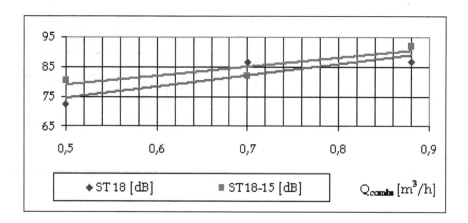

Fig. 24. Noise variation corresponding to ST 18-15 and ST 18 modules, depending on natural gas volume flow

No.	Natural gas flow rate Q_{combs} [m³/h]	Isotherms (according to the burning module type)	
		Module ST 18	Module ST 18-15
1	0,5		
2	0,7		
3	0,88		

Table 6. Isotherms at the infrared recording for modules ST 18 (left) and ST 18-15 (right) depending on the natural gas flow rate

7. Conclusions

The new generation of afterburning installations will need to respond to the performance requirements of the „smart" aggregates to automaticaly consider emissions, energetic efficiency and process requirements. The researches conducted at Suplacu de Barcau 2xST 18 Cogenerative Power Plant have analyzed the afterburning installation as integrated in the cogenerative group. Based on the measurements and the numerical results, the burning modules have been redesigned, experiments on the test bench have been conducted in order to establish the performances of the new generation of modular burners and the working regimes of the gas turbine have been established for water injection and afterburning cases. For the independent working of the TA-2 gas turbine, the numerical simulations had shown the possibility of 50% decrease in the NO_x emissions. However the modelling of the assembly TA-2 gas turbine with water injection and afterburning has shown that the NO_x emissions decrease (at the working regime defined by T_m = 1063 K) is possible only to 40 ppm (below 30 %). This confirms the necessity of a fine control of the quantity of water injected in the gas turbine particularly when a significant decrease in NO_x emissions is aimed. Future researches will involve test bench experimentations of the gas turbine working on natural gas with water injection, coupled with the multi-module afterburning installation. The experimental data should validate the elaborated numerical model and will constitute the design input data for a new afterburning installation.

8. Acknowledgments

The researches conducted at Suplacu de Barcau 2xST 18 Cogenerative Power Plant have been performed based on contracts 22-108/2008 and 21-056/2007 (Programme "Partnerships in priority fields") financed by Romanian Ministry of Education, Research, Youth and Sports. The consortium involved in the projects includes several Romanian companies: National Research and Development Institute for Gas Turbines COMOTI -

Bucharest, SC OMV PETROM SA, UPB CCT - Bucharest, SC OVM ICCPET SA - Bucharest, SC ICEMENERG SA - Bucharest, SC ERG SRL - Cluj, SC TERMOCAD SRL - Cluj.

9. References

Barbu, E., Rosu, I. & Turcu G. (2006). Supraincalzitorul cazanelor de la centrala cogenerativa 2xST 18 – Suplacu de Barcau, *ETCNEUR – 2006*, pp. 9-12, ISBN 973-7984-49-8, Bucuresti, 6-7 iulie 2006

Barbu, E., Ionescu, S., Vilag, V., Vilcu, C., Popescu, J., Ionescu, A., Petcu, R., Prisecaru, T., Pop E. & Toma T. (2010). Integrated analysis of afterburning in a gas turbine cogenerative power plant on gaseous fuel, *WSEAS Transaction on Environment and Development*, Vol. 6,. Issue 6, June 2010, pp. 405-416, ISSN 1790-5079

Barbu, E., Fetea, Gh., Petcu, R. & Vataman, I. (2010). *Arzator de postardere multimodular pe combustibil gazos*, Dosar OSIM nr. A/00999 din 21.10.2010

Benini, E., Pandolfo, S. & Zoppellari, S. (2009). Reduction of NO emissions in a turbojet combustor by direct water/steam injection: numerical and experimental assessment, *Applied Thermal Engineering*, doi: 10.1016/j.applthermaleng.2009.06.004

Carlanescu, C., Ursescu, D. & Manea, I. (1997). *Turbomotoare de aviatie – Aplicatii industriale*, Editura Didactica si Pedagogica, Bucuresti

Carlanescu, C., Manea, I., Ion, C. & Sterie, St. (1998). *Turbomotoare – Fenomenologia producerii si controlul noxelor*, Editura Academiei Tehnice Militare, Bucuresti

Conroy, J. (2003). Improving duct burner performance through maintenance and inspection, *Energy-Tech*, February 2003, http://www.energy-tech.com/article.cfm?id=17543

Ganapathy, V. (2001). Superheaters: design and performance, *Hydrocarbon processing*, July 2001, http://www.angelfire.com/md3/vganapathy/superhtr.pdf

McBride, B.J. & Gordon, S., (1992) *Computer Program for Calculating and Fitting Thermodynamic Functions*, NASA Lewis Research Center, NASA RP-1271, Cleveland, USA, http://www.grc.nasa.gov/WWW/CEAWeb/RP-1271.pdf

Neaga, C. (2005). *Tratat de generatoare de abur*. vol. III, Editura Printech, Bucuresti, ISBN 973-718-262-6

Petcu, R., (2010). *Contributii teoretice si experimentale la utilizarea gazului de depozit ca sursa de energie*, Teza de doctorat - Decizie Senat nr. 100/12.02.2010, Bucuresti

Popescu, J., Vilag V., Barbu, E., Silivestru, V. & Stanciu, V. (2009). Estimation and reduction of pollutant level on methane combustion in gas turbines, *Proceedings of the „3rd WSEAS International Conference on Waste Management, Water Pollution, Air Pollution, Indoor Climate – WWAI'09"*, pp. 447-451, ISBN 978-960-474-093-2, University of La Laguna, Tenerife, Canary Islands, Spain, July 1-3, 2009

Public Interest Energy Research, [PIER]. (2002). *Active control for reducing the formation of nitrogen oxides in industrial gas burners and stationary gas turbines*, California Energy Commission, http://www.energy.ca.gov/reports/2002-01-10_600-00-009.PDF

Zehe, M.J., Gordon, S. & McBride, B.J. (2002), *CAP: A Computer Code for Generating Tabular Thermodynamic Functions from NASA Lewis Coefficients*, NASA Glenn Research Center, NASA TP—2001-210959-REV1, Cleveland, Ohio, USA, http://www.grc.nasa.gov/WWW/CEAWeb/TP-2001-210959-REV1.pdf

8

Application of Statistical Methods for Gas Turbine Plant Operation Monitoring

Li Pan
Queen's University Belfast
U.K.

1. Introduction

Within a large modern combine cycle gas turbine (CCGT) power station, it is typical for thousands of process signals to be continually recorded and archived. This data may contain valuable information about plant operations. However, the large volume of data accompanied with inconsistencies within the data often limits the ability to identify useful information about the process. Utilising data mining techniques, such as Principal Component Analysis (PCA) and Partial Least Squares (PLS), it is possible to create a reduced order statistical model representing normal plant conditions. Such a model can then be utilised for fault identification and identifying possible improvements in key performance indicators such as thermal efficiency. Moreover, the gas turbine performance can be affected by changes in ambient conditions. A long term nonlinear PLS techniques can be applied here to investigate the seasonal changes in gas turbine.

In this chapter, an approach to establish a long term statistical model for gas turbine will be given, and the application of the model in fault detection and performance analysis will be demonstrated.

2. The data mining techniques

Within the power station, data are continuously collected and archived representing thousands of data points including temperatures, steam flow rates, pressures, etc. Potentially this data may contain valuable information about unit operation. However, collecting a large amount of data does not always equate to a large amount of information, leading to a lot of databases being regarded as data rich, but information poor. The task of extracting information from data is known as data mining, which is defined as the non-trivial process of identifying valid, novel, potentially useful, and ultimately understandable patterns in data. Data mining is the nontrivial process of extracting valid, previously unknown, comprehensible and useful information from large databases (Weiss and Indurkhya, 1998). Also, data mining is a generic term for a wide range of techniques which include intuitive, easily understood methods such as data visualisation to complex mathematical techniques based around neural networks and fuzzy logic (Wang, 1999; Olaru and Wehenkel, 1999). Applications are found within diverse areas such as marketing (Humby et al., 2003), finance (Blanco et al., 2002) and industrial process control (Martin et al., 1996). However, despite being a widely applied technique, it is reported that three quarters of all companies who attempt data mining projects fail to produce worthwhile results (Matthews, 1997). Unfortunately, this indicates that the potential of data mining techniques, with regard to the available data is often overestimated than the reality.

The act of data mining is itself part of a larger process known as knowledge discovery in data, KDD, which encompasses not only the analysis of data, but the gathering and preparation of data and the interpretation of results. Extracting knowledge from large data sets can be achieved through exploratory data analysis to discover useful patterns in data, in the form of relationships between variables.

Many techniques are applied as classification tools, to categorise new data following the analysis of a historical data set. In this chapter, the first method discussed in Section 2.1 is machine learning techniques which use a logical induction process to categorise a series of examples, resulting in decision tree and rules set which can be implemented in decision making processes. Typical application areas are fault diagnosis in industrial machines (Michalski et al., 1999) and the assessment of power system security (Voumvoulakis, 2010)

Case based reasoning methods as discussed in Section 2.2, are commonly applied to decision making tasks where previous experience is desirable, but may not be available. Case based reasoning provides an inexperienced user with exposure to experiences from others, through a set of historical 'cases', and has been of particular use in areas such as fault diagnosis (Wang et al. , 2008; Yan et al. , 2007) and system design and planning (Hinkle and Toomey, 1995).

Finally, Section 2.3 discusses multivariate statistical techniques, namely principal component analysis and partial least squares regression, which have been successfully applied to a range of applications areas including chemistry (Wold et al., 1987), manufacturing (Oliveira-Esquerre et al., 2004), and power system analysis (Prasad et al., 2007). It is also extended to finance (Blanco et al, 2002) and medicine (Chan et al, 2003) area. Principal component analysis and partial least squares regression are particularly popular in the area of chemometrics, where they are employed in the monitoring of processes which generate large and highly correlated data sets (Yoon and MacGregor, 2001; Kourti et al., 1996).

2.1 Machine learning

Machine learning techniques are those that use logical or binary operations to 'learn' a task from a series of examples, such as symptoms of medical or technical problems, leading to the diagnosis of the problem through the use of decision trees and rule sets which classify data using a sequence of logical steps (Michalski et al., 1999).

Decision trees are simple top down learning structures, which use Boolean classifiers to 'grow' a tree through recursive partitioning of the sample data using the available attributes. The development of a decision tree starts with the inclusion of all the training data in a root node, resulting in both correctly and incorrectly classified data. In order to 'grow' the tree, the data is recursively split by each attribute until all the attributes in the data have been used. Each node in the final tree, known as a leaf, represents a test on one of the attributes, and the branches from the node are labelled with the Boolean outcomes of the test (Quinlan, 1996).

The basic algorithm of building a decision tree, also known as ID3/C4.5 algorithm, follows a down rule (Quinlan, 1993). In the beginning, all the data is collected in the root node, and the data is recursively subdivided into fewer branches by assessing the information gain of each attribute in the training data to split the data further, until the terminal node which only contains one attribute is obtained (Quinlan, 1993).

Rule induction is achieved using a bottom up structure, starting with a rule that specifies a value for every available attribute on the decision tree, thereby making the rule as specific as possible. This rule is known as the seed and further rules are developed from it by successfully removing attributes one at a time, until more general rules are acquired. Any rule which includes a counter example is regarded as incorrect and is therefore discarded from the process. The rule learning terminates by saving a set of'shortest'rules. Also a new "RBDT-1" algorithm is devolved for learning a decision tree from a set of decision rules that cover the data instances rather than from the data instances themselves. The method's goal is to create on-demand a short and accurate decision tree from a stable or dynamically changing set of rules (Abdelhalim, A. 2009).

The primary advantage of decision trees is that their simplicity makes them very intuitive to users. However, large data sets call result in vast trees which can be 'needlessly' complex resulting in a largely unusable knowledge base : the ideal tree is as small and linear as possible. Due to their simple nature, decision trees are not suitable for more complex data structures and this is demonstrated by trees that, after pruning, still remain too large to be comprehensible.

2.2 Case based reasoning

Case based learning acquires knowledge from solutions to prior problems and employs it to derive solutions to the current problems. Once a current problem occurs, the similar case and previous solution are retrieved and possibly revised to better fit the current problem. The new solution can be retained into the case base in case to solve future problems. As a result, case based reasoning (CBR) systems are effectively used as lookup tables where 'the system' interrogates an indexed database of relevant cases, and one or more similar cases are retrieved and applied to discover an appropriate solution (Watson, 1999).

A significant issue in CBR is indexing, which limits the search space, thereby reducing case retrieval times. There are many methods for indexing, such as check list based indexing, which identifies predictive features for a case (inductive learning methods may be used) and places them on a list which is then used for indexing, and difference based indexing which selects features as indices that best differentiate one case from another. The user can also manually implement an indexing system, and it has been suggested that selection of indices by the user can be more effective than algorithms for practical applications (Kolodner, 1993). The indexed cases can then either be stored sequentially, making the system easy to maintain but slow to query for larger case sets, or using a hierarchical structure, which will organize cases so that only a small subset are considered during retrieval, thereby reducing search times (Smyth et al., 2001).

CBR is a self-maintaining system, the database of historical events is updated when new cases occurred and adding to the system's problem solving resources. The advantage of CBR is that it does not require a large number of historical data patterns to achieve satisfactory levels of performance : a CBR model may be created from a small number of cases and the case base can be refined over time (Hinkle and Toomey, 1995). CBR is particularly useful when studying data which has complex internal structures when there is little domain knowledge, enabling the sharing of experience.

Despite these benefits, CBR can be unsuitable for large scale applications as retrieval algorithms are inefficient when faced with handling thousands of cases. Maintenance of the case base, with respect to adding new cases and the removal of out date cases, may also be a problem as it is largely left to human intervention (Watson and Marir, 1994).

2.3 Multivariate statistical techniques

Statistical methods are employed to analyse the relationships between individual points in a data set, determining characteristics such as the average value and distribution of the data. The simple statistical measure represents a univariate approach to data analysis, which lacks the ability to constructively analyse large, multivariate data sets as the interactions between variables are ignored (Martin et al., 1996). In contrast, multivariate statistical analysis describes methods capable of observation and analysis of the multiple variables required for system monitoring (Kourti and MacGregor, 1995). This section discusses the multivariate techniques, principal component analysis (PCA), least squares regression and partial least squares (PLS), as they are more suitable to the analysis of large data sets than univariate methods.

Principal component analysis (PCA) is a statistical technique useful for identifying underlying systematic structures in data and separating it from noise (Wold et al., 1987). The identification of patterns in data structures allows PCA to be applied to problems requiring a reduction in the dimensionality of a data set, for example image processing(Bharati et al., 2003), or monitoring of industrial processes including chemical and microelectronics manufacturing processes (Wise and Ricker, 1991 ; MacGregor and Koutodia, 1995). These objectives are achieved by transforming variables, which are assumed to be correlated, into a smaller number of uncorrelated variables called principal components (PCs), providing a simpler description of the data structure. Each successive PC accounts for the most significant variability in the data in a particular direction, with the reduction in dimensionality achieved, the original data set can be represented by few PCs.

PCA is a useful tool when large data sets containing highly correlated variables are to be managed. PCA achieves reductions in data dimensionality, thereby simplifying future observation of variables : plotting a few PCs is significantly more convenient than plotting all original variables. Furthermore, the comparison capabilities between the historical information used to construct the model and newly presented data is a desirable characteristic for system monitoring applications (Martin et al., 1996). Fault identification can also be undertaken by analyzing the contribution of the independent variables to each PC (MacGregor et al., 1994).

Projection to latent structures, also known as partial least squares (PLS), is developed to solve the multi-collinearity problem in linear least squares regression (LSR) which can determine the best linear approximation for a set of data points (Freund and Willson, 1997). The benefit of PLS is achieved by identifying a set of uncorrelated, latent variables. This avoids the co-1inearity problems encountered by likelihood-ratio (LLR) test and utilises some of the techniques associated with PCA, with the new latent vectors composed of scores and orthogonal loading vectors. PLS regression is a robust, multivariate linear regression technique which is considered to be more suitable for the analysis and modelling of noisy and highly correlated data than LLR as parameters do not exhibit large variation when new data samples are included (Otto and Wegscheider, 1985). A high number of variables, with respect to the number of data samples, are also permissible in PLS, which can result in the modelling of noise for LLR(wise and Gallagher, 1996).

In summary, PLS is capable of producing robust, effective models, despite operational data limitations, for example, imprecise measurements and missing data (Oliveira-Esquerre,

2004). The ability to predict dependent data values, especially in the case of product quality data, which is often measured infrequently, is useful in process monitoring (MacGregor, 2005). The proficiency of PLS dealing with the highly correlated and collinear data is also frequently utilized in its application to process monitoring (Kresta, 1994).

2.4 Method selection

A selection of data mining techniques has been presented in this section and the characteristics of each shall now be considered with respect to the problems and available data presented by power plant monitoring and analysis.

Data derived from power plant monitoring can potentially consist of thousands of sensor measurements generated on a second by second basis. The data recorded is highly correlated, due to multiple sensors, which are in place to introduce redundancy into the measurements, and parallel paths within the system, for example, the steam and gas circuits. Data quality is also a factor, with noisy signals and missing data as the common problems. The correlated structure and data quality considerations present in power plant monitoring records bear strong similarities to other application areas discussed in this section, such as chemical process control(Ma et al., 2009; Ahvenlampi and Kortela, 2005), manufacturing (Oliveira-Esquerre et al., 2004; Baharati et al., 2004; Hisham et al., 2008) and medicine (Chan et al., 2003).

When monitoring a process which records a vast array of sensor data, individual analysis of each signal by a human operator is clearly not possible. The data analysis techniques discussed in this section can be applied to this problem as they identify essential correlations within the data. This simplifies the monitoring process by identifying the most relevant process signals and thereby reducing the search space.

When undertaking system monitoring, there are three main objectives. The first is the detection of a change in process operation and its nature - should this be a sensor fault, faults within the process or a change in product quality or system performance. Once a change in process behaviour is detected, diagnostic tests are then required to identify the cause of the change, which may require analysis of recent process data and/or consultation with an experienced operator. Finally, with the source of the change ascertained, a solution should then be identified and implementation of corrective action undertaken, if appropriate (Cinar and Undey, 1999). This may require either disabling the source of the problem, or in the event of a faulty sensor reading, reconstruction of the value.

The methods presented here provide different solutions to the process monitoring problem. Clustering, machine learning and CBR are diagnostic tools, which compare current fault conditions to historical examples of faults. While these techniques will identify the nature of the problem, they are not capable of detecting its occurrence or providing signal reconstruction in the event of a sensor fault. Statistical methods offer a model based approach, where process operation data is compared to a system model, based on 'normal' operating conditions. This provides continuous process monitoring, which can supply early warning of a small process change and offers operators the opportunity to take action to prevent the fault becoming more serious. In particular, PCA and PLS can supply the operator with information as to which process variables are outside normal limits (MacGregor et al., 1994). This serves to focus the operators' attention on the problem area, allowing process knowledge to be used to identify the source of the problem.

Not all of the techniques discussed in this section are capable of providing a solution that enables preventative action to be taken for CCGT power plant.

CBR, providing a historical event that is similar to the currently observed event, is also of little use in this instance, as it again offers classification of new events, however, an explanation of the similarities identified between cases is unavailable.

Cluster analysis and 'rules' may be useful in identifying groups of similar events, as their aim is to suggest correlations in data. Once similarities have been identified for events resulting in groupings with common characteristics further investigation would be required to identify the relationships between variables that cause these groupings. A more complete solution is offered by statistical methods.

Both PCA and PLS are capable of identifying correlations within data, while PLS also offers the ability to extend this to identifying the correlations which are predictive of a dependent quantity. The correlations identified within the data can then be studied using the scores and loading vectors obtained, indicating the contribution of variables, if any, to the variation of the dependent parameter. The historical data is suitable for the development of a system model, by means of PCA and PLS, which can be applied to continuous monitoring.

Archived data is available, detailing sensor data, for example temperatures, pressures, etc, throughout the plant at regular intervals. Once the PCA and PLS models are developed, it can provide a relatively straight forward model which has both the ability for online fault monitoring and offline performance analysis. For practical application, those PCA and PLS models are required for fast online response within 20 seconds, and reasonable prediction accuracy in a wide operation range.

With PCA and PLS identified as possessing properties that are useful in relation to the problems posed by power plant and power system operation, the statistical modelling methods will provide the most suitable approach for operation monitoring and performance analysis of CCGT power station.

3. PCA and PLS algorithm

Give an original data matrix X $(m \times n)$ formed from m samples of n sensors, and subsequently normalised to zero mean and unit variance, can be decomposed as follows:

$$X = TP^T + E = t_1 p_1^T + t_2 p_2^T + ... + E = PC_1 + PC_2 + ... + E \qquad (1)$$

where $T \in R^{m \times A}$ and $P \in R^{n \times A}$ are the principal component score and loading matrices, E is the residual matrix (Lewin, 1995).

The principal component matrices can be obtained by calculating eigenvectors of original data. Following the creation of the correlation matrix of original data, the corresponding eigenvalues and eigenvectors are calculated, where an eigenvalue is an eigenvector's scaling factor. As the eigenvectors with the largest eigenvalues correspond to the dimensions than have the strongest correlation in the dataset, the data can then be ordered by eigenvalue, highest to lowest to give the components in order of significance (Jolliffe, 2002). There are a number of methods available to determine the number of ordered PCs. A cross validation which calculates the predicted error sum of squares (PRESS) (Valle et a., 1999) is provide more reliable solutions than a simple scree test (Jackson, 1993).

Partial least square requires two block of data, an X block (input variables) and Y block (dependent variables). PLS attempts to provide an estimate of Y using the X data, in a similar manner to principal components analysis (PCA). If T and U represent the score matrixes for the X and Y blocks, and P and Q are the respective loadings, the decomposition equations can be presented as:-

$$X = TP^T + E \qquad (2)$$

$$Y = UQ^T + F \qquad (3)$$

where E and F are the residual matrices. If the relationship between X and Y is assumed to be linear then the residual matrices E and F will be sufficiently small, and the score matrices T and U can be linked by a diagonal matrix B such that:

$$U = BT \qquad (4)$$

Hence the predicted dependent variable can be translated (Flynn, 2003) as:

$$\hat{Y} = BTQ^T + E + F \qquad (5)$$

4. Nonlinear modeling approach

As we discussed previously, PCA and PLS model are powerful linear regression techniques. However, in the real power generation industry, many processes are inherently nonlinear. When applying linear model to a nonlinear problem, the minor latent variables cannot always be discarded, since they may not only describe noise or negligible variance structures in the data, but may actually contain significant information about the nonlinearities. This indicates that the linear model may require too many components to be practicable for monitoring or analyzing the system.

Recognition of the nonlinearities can be achieved using intuitive methods, for example, which apply nonlinear transformations to the original variables or create an array of linear models spanning the whole operating range. More advanced methods have also been proposed including nonlinear extensions to PCA (Li et al. 2000), introducing nonlinear modifications to the relationship between the X and Y blocks in PLS (Baffi et al., 1999) or applying neural network, fuzzy logic, etc. methods to represent the nonlinear directly.

Transformation of the original variables using nonlinear functions can be introduced prior to a linear PCA and PLS model. For this purpose, the input matrix X is extended by including nonlinear combinations of the original variables. However, process knowledge and experience is required to intelligently select suitable nonlinear transformations, and those transforming functions must sufficiently reflect the underlying nonlinear relationships within the power plant. Another problem with this approach is the assumption that the original sets of variables are themselves independent. This is rarely true in practice, which can make the resulting output from the data mining exercise difficult to interpret.

An alternative and more structured approach is the kernel algorithm. The purpose of kernel algorithm is to transform the nonlinear input data set into a subspace with kernel function. In the kernel subspace, the nonlinear relationship between input variables can be transformed into linear relationship approximately. By optimising the coefficients of kernel function, the transformed data can be represented using a Gaussian distribution around linear fitting curve in the subspace. Furthermore, introducing neural network approaches into the kernel structure is generally seen to be more capable of providing an accurate representation of the relationship for each component (Sebzalli and Wang, 2001). In this area, the multilayer perceptron (MLP) networks are popular for many applications. However the initial model training is a nonlinear optimization problem, requiring conjugate gradient and Hessian-based methods to avoid difficulties arising from convergence on local minima. In order to solve this problem, a radial basis function (RBF) network has been selected over other approaches, due to its capability of universal approximation, strong power for input and output translating and better clustering function. A standard RBF network consists of a single-layer feedforward architecture, with the neurons in the hidden layer generating a set of basis functions which are then combined by a linear output neuron. Each basis function is centered at some point in the input space and its output is a function of the distance of the inputs to the centre. The function width should be selected carefully because each neuron should be viewed to approximate a small region of the input surface neighboring its centre. Therefore, the RBF network also has been named localized receptive field network. This localized receptive character implies a concept of distance, e.g. the RBF function is only activated when the input has closed to the RBF network receptive field. For this reason, the performance of RBF network is more dependent on the optimisation of RBF function coefficients rather than the type of function (Jiang et al., 2007).

In order to reduce the neural network dimension, the input data are firstly decomposed into few components, then the output can be reconstructed with nonlinear relationship. Hence, each component will possess its own nonlinear function $f_{non-linear}$, so that

$$\hat{X} = f_{non-linear}^{X}(T) \tag{6}$$

$$\hat{Y} = f_{non-linear}^{Y}(T) \tag{7}$$

In this research, radial basis functions have been selected to represent the non-linearities, since once the RBF centres and widths have been chosen, as described below, the remaining weights can be obtained using linear methods.

4.1 RBF network
The radial basis function network employed in this research is illustrated in Figure 1.

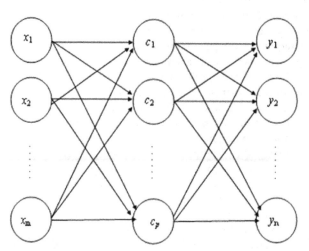

Fig. 1. Radial basis function network

The network topology consists of m inputs, p hidden nodes and n outputs, and the network output, y_i, can be formulated as:-

$$y_i = \sum_{j=1}^{p} w_j^{(i)} \theta_j \left(\left\| \mathbf{X} - \mathbf{c}_j \right\| \right) \quad i = 1, 2, \dots n \tag{8}$$

where, $w_j^{(i)}$ are weighting coefficients, and θ_j is the basis function. In this research, a Gaussian base function was selected, which is defined as:-

$$\theta_j \left(\left\| \mathbf{X} - \mathbf{c}_j \right\| \right) = \exp \left[-\sum_{k=1}^{m} \left(\frac{x_k - \mathbf{c}_j}{\sigma_j} \right)^2 \right], \quad i = 1, 2, \dots p \tag{9}$$

The Euclidean distance $\left\| \mathbf{X} - \mathbf{c}_j \right\|$ represents the distance between the input space $\mathbf{X}$ and each RBF centre $\mathbf{c}_j$, where $\mathbf{X} = [x_1 \ x_2 \ \dots \ x_m]$, and σ_j is the width coefficient of each RBF node. The coefficient matrix $[\sigma, c, w]$ is obtained off-line using a suitable training algorithm. Some of the more popular options are least mean squares (LMS) (Moody et al., 1989), orthogonal least squares (OLS) (Li et al,. 2006) and dual-OLS (Billing et al., 1998). These traditional algorithms often employ a gradient descent method, which tends to converge on local minima. In order to address the global optimisation problem, a recursive hybrid genetic algorithm (RHGA) (Li and Liu, 2002, Pan et al., 2007) is employed here to search for valid solutions.

4.2 The genetic algorithm

The typical genetic algorithm (GA) is based upon *survival of the fittest*, and the network framework $[\sigma, c]$ is coded into the binary genes as illustrated in Table 1. The initial population are selected at random from the entire solution space, with the binary coding denoting whether the training samples are selected as the centers of the hidden neurons (Goldberg, 1989).

All the potential hidden centers	A randomly created gene code	Coded network framework
$[\sigma_1, \hat{c}_1]$	1	$[\sigma_1, \hat{c}_1]$
$[\sigma_2, \hat{c}_2]$	0	---
$[\sigma_3, \hat{c}_3]$	0	---
$[\sigma_4, \hat{c}_4]$	1	$[\sigma_4, \hat{c}_4]$
$[\sigma_5, \hat{c}_5]$	1	$[\sigma_5, \hat{c}_5]$

Table 1. Encoding scheme of genes

For each generation, random crossover and mutation is applied to the genes, leading to a new generation of network frameworks being obtained. The fitness, f, of the new population is determined using:-

$$\frac{1}{f} = \sum_{j=1}^{n} (\hat{y}_j - y_j)^2 \tag{10}$$

where, $\hat{y}_j$ is the jth RBF output and y_j is the actual value. The most recent framework will be retained if its fitness improves upon previous generations.

Although the genetic algorithm has the capability of wide region searching and efficient global optimizing, it is weak in some local point fitting. This may lead to a decrease in model accuracy. Therefore, the genetic and gradient descent algorithm can be combined in order to obtain both the global and localize optimizing capability (Pan, et al., 2007). In this hybrid algorithm, an initial optimized network can be obtained by the genetic algorithm, and then the structure of network can be further shaped for some specific points with the gradient descent algorithm. The next step is to examine the variate of fitness coefficient. If the fitness reached the preset bound then the regression will be completed, otherwise, the network will be reconstructed for next generation optimisation, and repeat the gradient descent regression, until reach the preset number of generations or meet the request fitness.

5. The auxiliary methods

Once a PCA/PLS model for normal operating conditions has been developed, the real time online DCS data then can be applied into the model to obtain a reconstruction of input data. It can be used to determine whether recorded plant measurements are consistent with historical values and neighboring sensors. A comparison can then be made between the reconstructed value for each variable and the actual measurements. Performed manually this can be a time consuming task. In this section, some efficient auxiliary methods will be discussed for the quality control, sample distribution analysis and fault identification.

5.1 Quality control method

There are two approaches that can quickly help to identify differences between the actual and reconstructed value of a variable, which are the squared prediction error (SPE) and Hotelling's T^2 test.

The SPE value, also know as the distance to the model, is obtained by calculating a reconstruction of each variable, $\hat{x}_i$, from the model, and then comparing it with the actual value, x_i. The SPE for all variables in each data sample can be calculated as

$$SPE = \sum_{i=1}^{n}(x_i - \hat{x}_i)^2 \tag{11}$$

In order to distinguish between normal and high values of SPE, a confidence limit, known as the Q statistic test is available, which can be determined for α percentile confidence as:

$$Q_\alpha = \theta_1 \left(\frac{c_\alpha \sqrt{2\theta_2 h_0^2}}{\theta_1} + 1 + \frac{\theta_2 h_0 (h_0 - 1)}{\theta_1^2} \right)^{1/h_0} \tag{12}$$

where c_α is the confidence coefficient for the $1-\alpha$ percentile of a Gaussian distribution, θ_i is the sum of unused eigenvalues to the ith power and h_0 is a combination of θ as outlined below:

$$h_0 = 1 - \frac{2\theta_1 \theta_3}{3\theta_2^2} \tag{13}$$

The T² statistic test is designed as a multivariate counterpart to the student's t statistic. This test is a measure of the variation within normal operating conditions. With Tracy- Widom distribution, the T² test can be extended to detect peculiar points in the PCA model (Tracy et al., 1993).

Given h components in use, t_i is the ith component score and s_i is its covariance, then the T² can be defined as

$$T^2 = \sum_{i=1}^{h} \frac{t_i^2}{s_i^2} \tag{14}$$

As with SPE, an upper control limit, T_α^2 can be calculated with n training data. This relates the degrees of freedom in the model to the F distribution,

$$T_\alpha^2 = \frac{h(n^2 - 1)}{n^2(n - h)} F_{1-\alpha}(h, n - h) \tag{15}$$

It should be noted that a rise in the SPE or T² value does not always indicate a fault, it also may be caused by the process is moving to a new event which is not accounted in the training data. Additionally, both indicators are affected by noise on the system and deviation of measurements from a normal distribution. This can result in nuisance values for both SPE and T². However, false alarms can be largely eliminated by simple filtering, and adjustment of the associated threshold (Qin et al., 1997).

5.2 Sample distribution
Both the SPE and T² are unlikely to differentiate between a failing sensor and a fault on the power plant. In this case, a plotting of t scores can be combined with the previous methods to distinguish between the two conditions.

The PCA model gives a reduction of data dimension with minimum information less. Therefore, the original m dimension data can be plotted in a plane coordinated by the first two components, and the relative position between each data point is remained the same as the original m dimension space. This character gives a capability to directly observe the similar distribution structure of original sample data, in a 2-dimension plane.

Especially, quoting the T² control limit into the 2-dimension plane, we have

$$T_{2-D}^2 \leq T_{\alpha\,2-D}^2 \tag{16}$$

substituting Eq. (14) and (15), the Eq. (16) can be transformed as

$$\left(\frac{t_1^2}{s_1^2} + \frac{t_2^2}{s_2^2} \right) \leq \frac{2(n^2 - 1)}{n^2(n - 2)} F_{1-\alpha}(2, n - 2) \tag{17}$$

Define

$$c = \frac{2(n^2 - 1)}{n^2(n - 2)} F_{1-\alpha}(2, n - 2) \tag{18}$$

then it gives that

$$\frac{t_1^2}{s_1^2} + \frac{t_2^2}{s_2^2} \le c \tag{19}$$

Eq. (19) defines a control ellipse for t-score plotting. Score for normal operating conditions should fall within this ellipse. So when a process fault occurs, the individual points on the t score plots may be observed drifting away from the normal range into a separate cluster. The relative position of these fault clusters can assist in latter diagnosis.

5.3 Fault orientation

Having confirmed that there is a sensor fault, and not a process condition, the next step is to identify which sensor is failing. If a signal is faulty, a significant reduction in SPE before and after reconstruction would be expected. However, in practice the reduction in SPE can affect all inputs, making the faulty sensor unidentifiable. This situation arises due to a lack of redundancy, or degrees of freedom, among the measurements.

The above difficulties can be overcome by calculating a sensor validity index (SVI) (Dunia et al, 1996). This indicator is determining the contribution of each variable to the SPE value. The SPE value should be significantly reduced by using the reconstruction to replace the faulty input variable. If an adjusted data set z_i represents a input set with the x_i variable being replaced by reconstructed data $\hat{x}_i$, and the adjusted model predicted value being $\hat{z}_i$, then the sensor validity index for i^{th} sensor η_i can be defined as

$$\eta_i^2 = \frac{(z_i - \hat{z}_i)^2}{\textbf{SPE}} \tag{20}$$

The SVI is determined for each variable, with a value between 0 and 1 regardless of the number of samples, variables, etc. The value of SVI close to unity is indicative of a normal signal, while a value approaching zero signifies a fault. It is assumed that a single sensor has failed, and the remaining signals are used for reconstruction. Also, system transients and measurement noise can lead to oscillations in SVI, and possibility of false triggering. Consequently, each signal should be filtered and compared with a user-defined threshold.

6. Application of PCA and PLS model

As these power plants operate in a competitive market place, achieving optimum plant performance is essential. The first task in improving plant operation is the enhancement of power plant operating range. This power plant availability is a function of the frequency of system faults and the associated downtime required for their repair (Lindsley, 2000). As such, availability can be improved through monitoring of the system, enabling early detection of faults. This therefore allows the system working at non-rated conditions, corrective actions, or efficient scheduling of system downtime for maintenance (Armor, 2003).

Monitoring of power plant operations is clearly an important task both in terms of identifying equipment faults, pipe leaks, etc. within the generating units or confirming sensor failures, control saturation, etc. At a higher level, issues surrounding thermal efficiency and emissions production for each generating unit, as measures of plant performance, and the seasonal influence of ambient conditions will also be of interest. Fortunately, the frequency of measurement and distribution of sensors throughout a power tation provides a great deal of redundancy which can be exploited for both fault identification and performance monitoring (Flynn et al., 2006). However, modern distributed control systems (DCSs) have the ability to monitor tens of thousands of process signals in real time, such that the volume of data collected can often obscure any information or patterns hidden within.

Physical or empirical mathematical models can be developed to describe the properties of individual processes. However, there is an assumption that faults are known and have been incorporated into the model. This can be a time-consuming exercise and requires the designer to have extensive knowledge of the application in question (Yoon and MacGregor, 2000). Alternatively, data mining is a generic term for a wide variety of techniques which aim to identify novel, potentially useful and ultimately understandable patterns in data. The most successful applications have been in the fields of scientific research and industrial process monitoring, e.g. chemical engineering and chemometrics (Ruiz-Jimenez et al., 2004), industrial process control (Sebzalli et al., 2000) and power system applications such as fault protection in transmission networks (Vazquez-Martinez, 2003). In the following sections it will be shown how using the principal component analysis (PCA) technique. It is possible to exploit data redundancy for fault detection and signal replacement, as applied to monitoring of a combined cycle gas turbine.

Furthermore, the archived data is used to assess system performance with respect to emissions and thermal efficiency using a partial least square (PLS) technique.

5.1 Raw data pre-process

The PCA and PLS models are trained using historical data to suit the 'normal' plant operating, and the training data have to be selected carefully to avoid failing and over range data from normal power plant operation. The normal power plant operation was defined around the typical output range of 60 MW – 106 MW for single shaft unit and 300 MW – 500 MW for multi-shaft unit. There are severe dynamic conditions existing in the starting up and shutting down period. Therefore, those periods has to be removed from raw data archives. An instance is illustrated in Figure 2, for a single shaft unit operation, approximately one hour operating data was removed after and before system shut down and start up, in order to avoid the transient process.

The DCS normally collects sensor data every second, however, due to the power plant parameters are mainly consisted by temperature and pressure signals, the typical power plant responding time is around minutes. Therefore, consider of the balance of computational complexity and information quality, the sampling interval was determined as 1 minute. Since the raw data sample was archived from DCS, it still contains lots of anomalous signals such as break down process, which the power out suddenly crash down. Noised signal, is a signal disturbed by white noise. And spike, is an instantaneous disturbance which can cause a far deviation from normal signal level. Those data must be pre-filtered before being employed to train a model.

It is generally recognized that CCGT performance, and in particular gas turbine performance, can be affected by changes in ambient conditions (Lalor and O'Malley, 2003). For example, a fall in barometric pressure causes a reduction in air density and hence inlet compressor air flow. Similarly, an increase in ambient temperature causes a reduction in air density and inlet compressor air flow. Since the turbine inlet temperature is maintained as a constant, there is a subsequent reduction in turbine inlet pressure and hence cycle efficiency.

Variations in other external variables such as relative air humidity and system frequency (affecting compressor rotational speed) can also impact on gas turbine performance. Therefore, the training data selection for a widely suitable PCA model has to contain the information of the seasonally changes of ambient condition.

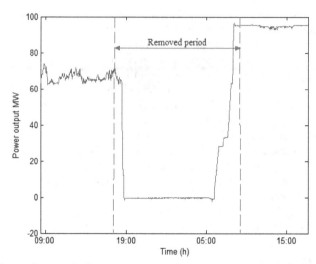

Fig. 2. Removed transient period

In order to obtain a entire seasonal model, the training data sorting process is designed to archive power plant operating data for years, then split all of the ambient variables into many small intervals, and pick up a sample data from each intervals to ensure that the training data contain the operating information for every ambient conditions.

6.2 Sensor data validation

With aging sensors, and the associated performance degradation, inevitable, faulty sensors are a relatively common occurrence in system monitoring. A common example of sensor failure is 'stuck at' signal, as illustrated in Figure 3 (a), which the fault is occurred at 300th data point. The following data is missed and the sensor's output is stuck at the last measurement. Another example is drifting signal, shown as Figure 3 (b), that the original data is disturbed by an increasing interference. Also, a biased signal is a constant noise which biased the sensor's data to other level, as shown in Figure 3 (c).

Univariate limits, i.e. upper and lower bounds are often applied to the detection of these faults. Problems such as biased sensors can be detected when the value eventually exceeds the predefined limits. However, a faulty signal within the univariate limits, such as a drifting sensor, will often go undetected for a long period of time. In order to identify such those faulty sensors, a multivariate approach is required, which will give consideration to the sensor value as part of wider plant operation.

Furthermore, if a sensor is faulty, an operator may choose to disable the sensor, but if the signal is used for feedback/feedforward control, disabling the sensor can only be part of the solution. In this instance, the problem can normally be resolved by signal reconstruction based upon sensor readings from neighboring sensors in the plant. This will require a system model, operating in parallel with the real plant.

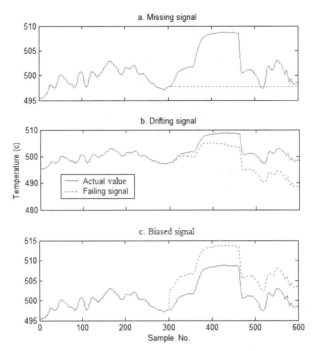

Fig. 3. Sensor faults

Principal component analysis, PCA, as a suitable technique for sensor monitoring and validation as it captures data variability for normal process operation. The development of a PCA model is intended to reduce the dimensionality of a set of related variables, while retaining as much of their variance as possible. This is achieved by identifying new, latent variables known as principal components, PCs, which are linearly independent. A reduced set of these latent variables are then used for process monitoring, with a small number of components normally sufficient to capture the majority of variability within the data.

Monitoring of a system using PCA is a modeling based approach, achieved by comparing observed power plant operation to that simulated by the model from available sensor data. The comparison between model and plant data, resulting in residuals, can then determine if the recorded information is consistent with historical operation and neighboring sensors. Faults are detected by observing deviations from normal operation, which can then be investigated to determine the exact source of the problem.

There are two common automated methods to compare recorded data with the model, as defined in section 5.1, the squared prediction error, SPE, and Hotelling's T2 test. Also, the sensor validity index, SVI, will identify failing sensors, and t score plots, from a cluster representing normal, fault free operation. All of those techniques are detailed in section 5. If an individual sensor is identified as being at fault, it can be replaced with a value reconstructed by the PCA model from other sensor data. However, if the fault is actually with the power plant, corrective maintenance or other necessary action should then be scheduled.

6.3 PCA model performance

In order to demonstrate the monitoring capabilities of this PCA model, a drift signal is introduced to the testing data set. As shown in Figure 4, the drift occurred in the sensor monitoring the steam temperature at 5:00 am. Generally, the lower bound of steam temperature is 500 °C during power plant normal operating period. Consequently, this drift can be detected by under limit indicator approximately 2 hours after the drift was introduced. In contrast to sensors limit indicator, the associated squared prediction error (SPE) monitoring test is illustrated in Figure 5 and shows that the SPE test detects the sensor fault 30 minutes after the introduction of the drift with 95% confidence limit, and 45 minutes with 99% confidence limit. Similarly, the T-squared test detected this sensor fails within 35 minutes using 95% confidence limit, and it crossed the 99% threshold 10 minutes later, as shown in Figure 6. The earlier SPE and T-square fault identification can provide more time for the power plant operator to take actions to solve problems.

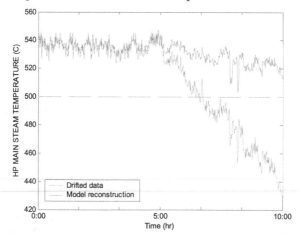

Fig. 4. Sensor drifts for single shaft CCGT unit

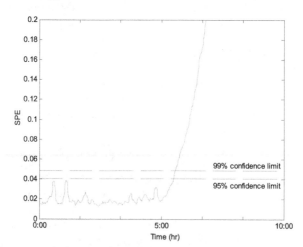

Fig. 5. SPE test for sensor drift in single shaft CCGT unit

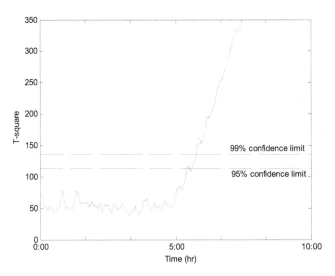

Fig. 6. T-square test for sensor drift in single shaft CCGT unit

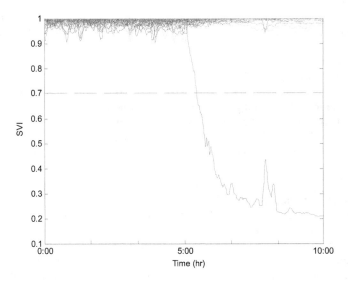

Fig. 7. SVI for sensor drift in single shaft CCGT unit

Following the detection of sensor fault condition, the source of the problem must be identified. Calculation of the sensor validity index (SVI), described in section 5.3, the variations in the SVI for each sensor are illustrated in Figure 7. According to the defined threshold of 0.7, the SVI chart clear identified the faulted sensor at 5:40 am, with the associated index of this signal falling into the range 0.7 to 0.2. Also, system transients and measurement noise can lead to oscillations into the SVI and there is a clearly example of SVI oscillations caused by system transient during 8:00 am to 9:00 am. It should be noted that when the HP main steam temperature signal drifts, the associated indices for the remaining sensors rise toward unity, accentuating identification of the biased sensor. As the fault is with the sensor, not the process, the PCA model can undertake reconstruction of the failed sensor as shown in Figure 4.

6.4 PLS model performance

Having validated the observed sensor data with PCA processor, optimisation of power plant performance should now be addressed. In order to maximize the power generation and simultaneously minimize the fuel consumed and pollution emissions, the performance variables must be able to monitor online and the internal relationship between the performance variables and associated operating parameters should be able to examine through offline analysis.

However, recover some performance variables is a multi-dimensional problem, such as the thermal efficiency, which depends on power demand, supplied fuel type, even the ambient conditions. Due to the expense and complexity of performance variables monitoring, development of online performance monitoring, capable of determine power plant performance from a variety of process variables, is often desirable. Validated and archived plant data can be employed to develop models which are capable of predicting the quality of process operation while providing an insight into the relationship between quality and associated process conditions.

PLS as a suitable technique for plant monitoring and shall be implemented here to demonstrate how system data can be applied to obtain a model of normal plant operation, with respect to a variety of quality variable measures, such as power plant efficiency, emissions and so on.

As with PCA, monitoring of individual fault conditions is not necessary and problems are instead detected as deviations from normal operation. With load cycling of generation plant increasingly common, a wide range of operating conditions are detailed in archived plant data and potentially contain indicators of operating conditions which lead to optimal power plant performance. The availability of operator logs makes it possible to indentify period of generation regarded by operators to be representative of fault-free power plant performance.

6.4.1 Variance explanation contribution

A benefit from the PLS model is that it has the ability to examine the effect of each input variable on the quality variables. Since the PLS model determines the variance explanation contribution of each variable by examining the correlation to the output variables, the PLS model is not only able to find those variables which have the greatest effect on output, but also can find the variables have indirect effect on the quality variables. This function can be applied to research the effect of any variable we interested, such as air temperature, sea water temperature, humidity and so on.

For instance, the variable contributions to the variance explanation of efficiency are charted in Figure 8 for a normal CCGT plant. Since the input variables are selected for highly related to the efficiency, most of them have comparatively high value of variance explanation, and these can be considered to be important variables to be monitored and/or adjusted when attempting to achieve enhanced operating goals. The most important variables are varying similar in both single shaft and multi shaft unit. For example, it can be observed that the No.1 and 2 are outstanding with 87.2% and 86.8% variance explained in the single shaft model, and they pointed to the signals of power output and gas flue flow respectively. Contrasts to multi-shaft model, above variables are identified as parameter No.1 with 85.0% explanation for power output and No.6 with 83% explanation for gas flue flow. Also, a group of sensors measuring the high pressure steam parameters are significant, which is the No.19-21 in the single shaft model with around 85% contributions and No. 27-29, 60-62 for both gas turbines in the multi-shaft model with around 80% contributions.

In addition, variations in ambient conditions is also interested, the last 4 variables in both models represent the effects of humidity, air temperature, barometric pressure and sea water temperature, respectively. It is significant that the sea water temperature has an extremely high effect on the power plant efficiency. The reason is considered of the condensing with sea water. The cooler sea water increases heat transfer from the condensing steam, and hence increase the thermal efficiency.

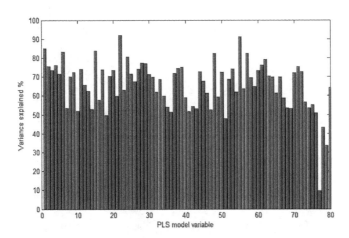

Fig. 8. Variance explanation for CCGT efficiency in multi-shaft unit

6.4.2 Relationship curve

From previous section, the PLS variance explanation suggest that sea water temperature is the most significant ambient condition for thermal efficiency. In order to better appreciate the impact of these environmental variables on the model, we introduce a new technique to study the relationship between input and output variables. It is of interest to lock all model inputs at a normal operating point, e.g. the power output is 90 MW, IGV position is 77% and etc., except the ambient variable being considered. For the simpler structure and closer variables relationship, the instance is chosen to use the single shaft unit, and consequently, Figure 9 illustrate the relative impact of these input parameters on the associated quality output measure for the CCGT plant.

It can be seen that increasing sea water temperature can significantly reduce the efficiency, the linear curve shows that about 50% increase in sea water temperature can cause 8% decrease in efficiency. Observably, the nonlinear curve shows that the relationship between ambient conditions and efficiency is more complicated and non-monotonic. As shown in the red line in Figure 9, the effects of ambient air and sea water temperature on the plant efficiency are represented as two tendency directions. One is the temperature upon the 12 degree, the efficiency is decreased following the increase in the environment temperature, and the reason is well known as we discussed in section 6.1 that the reduced cooling water temperature can enhance the steam cycle efficiency. Another direction present an interesting result where the temperature is lower than 12 degree, the efficiency enhancement seems to be more difficult when the temperature goes down. The reason is considered that the air and fuels inlet temperature will be excessively decreased during chill period and causes the decrease in gas combustion temperature and consequently reduced the gas turbine efficiency; it tends to counteract the effects of decrease in cooling water temperature on efficiency enhancement.

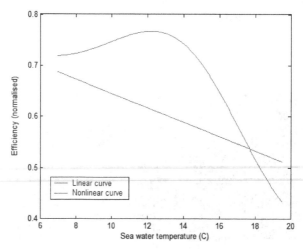

Fig. 9. Relation curve for single-shaft unit: efficiency vs. sea water temperature

7. Conclusion

Distributed control systems provide many advantages in terms of improvements in productivity and plant manoeuvrability when introduced into power plants and other industrial processes. However, the ease of access to a range of plant-wide signals potentially introduces vast problems of scale, since the meaningful information contained within the collected data may be somewhat less than the volume suggests. The task remains, therefore, to identify normal operating regions and relationships within the historical data, and subsequently to apply the collated rules, reference cases, etc. Principal component analysis has received considerable interest as a method of reducing the effective measurement space, and has been considered here for process monitoring of a combined cycle gas turbine.

Traditionally, operator practice has been reactive, whereby actions are taken following the triggering of process alarms, often set over-responsive and mode insensitive – PCA methods enable a more proactive role for the operator, providing early warning of plant irregularities, and identification of instrumentation errors and process faults. The PCA model is identified under normal operating conditions, and subsequently unusual deviations are highlighted and identified.

On the other hand, The PLS model has received considerable interest as a method of analyzing process data and in this instance it has been used for analyzing a combined cycle gas turbine. Analysis of the models variance demonstrates that CCGT performance is affected by changes in ambient conditions. Also a relation curve method can be utilized here to study the impact of these external parameters, from the environment, have on the gas turbine performance.

Future work could extent the quality measurements (efficiency and emissions) to include other important requirements such as plant life, unit flexibility or cost of generation. Where the objective is to optimize power plant quality measurements and consequently enhance plant performance.

Furthermore, it is considered to integrate the physical or empirical mathematical models with the statistical model. Since the statistical models are limited by training range, it is unreliable to be employed under untrained conditions. Compared to physical model, a trained statistical model is difficult to tune in case of power plant renovation or update, due to it requests the input variables remaining a designed internal relationship.

On the other hand, developing the pure physical model can be a time-consuming exercise and requires the designer to have extensive knowledge of the application. Especially for the large scale system, the relationship between each variable is highly correlated and tangled, it is impossible to build a physical model which can provide a detailed and accurate representation of the operation. Therefore, utilizing the statistical model to substitute the complex components in a physical model could be good solution for reduce modeling calculation. Similarly, the adjustability and flexibility of statistical model can be increased by integrating some physical control loop or simulation into the statistical model.

8. References

Abdelhail, A., I. Traore and B. Sayed: 'RBDT-1: A New Rule-Based Decision Tree Generation Technique', *Rule Interchange and Applications*, 58, pp. 108-121, 2009

Ahvenlampi, T. and U. Kortela:' Clustering algorithms in process monitoring and control application to continuous digesters', *Informatica*, 29, pp.101-109, 2005

Armor, A. F.: 'Management and integration of power plant operations', *Thermal power plant simulation and control*, IEE Power and Energy Series 43, pp. 395-416, 2003

Baffi, G., E. B. Martin and A. J. Morris: 'Non-linear projection to latent structures revisited: the quadratic PLS algorithm', *Computers and Chemical Engineering*, 23, (3), pp. 395-411, 1999

Billings S.A and Hong, X.: "Dual-orthogonal radial basis function networks for nonlinear time series prediction". *Neural Networks*, 11 (3), pp. 479-493, April 1998

Blanco, C., D. Soronow and P. Stefiszyn: 'Multi-factor models for forward curve analysis: An introduction to principal component analysis', *Commodities Now*, June, pp. 76-78, 2002

Bharati M. H., J. J. Liu and J. F. MacGregor: 'Image texture analysis: methods and comparisons', *Chemometrics and Intelligent Laboratory Systems*, 72, PP. 57-71, 2004

Chan, C.: 'Application of multivariate analysis to Optimize Function of Cultured Hepatocytes', *Biotechnology Progress*, 19, (2), pp. 580-598, 2003

Cinar, A., and C. Undey: 'Statistical process and controller performance monitoring: A tutorial on current methods and future directions', *Proceedings of the American Control Conference*, San Diego, USA, 4, pp. 2625-2639, 1999

Dunia, R., S. Qin, T.F. Edgar and T.J. MCAVOY: 'Identification of faulty sensors using principal component analysis', *AIChE Journal*, 42, (10), pp. 2797-2812, 1996

Flynn, D., J. Ritchie, M. Cregan and L. Pan: 'Data mining techniques applied to power plant performance monitoring', *16th IFAC World Congress, Prague*, 2006.

Freund, R.J., and W.J. Wilson: 'Regression analysis: statistical modelling of a response variable', Academic Press Ltd., London, UK, 1997

Goldberg, E.D.: " Genetic Algorithms in Search, Optimization, and Machine Learning". *Addision-wesley publishing company*, INC. pp. 27-95, 1989

Jackson, D.A.: 'Stopping rules in principal component analysis: a comparison of statistical and heuristical approaches', *Ecology*, 74, (8), pp. 2204-2214, 1993

Jackson, J.E.: 'A user's guide to principal component analysis', Wiley series in probability and statistics, John Wiley & Sons, Hoboken, New Jersey, 2003

Jolliffe, I.T.: 'Principal component analysis', Springer series in statistics, New York, 2002

Hinkle, D. and C. Toomey: 'Applying case-based reasoning to manufacturing', *AI Magazine*, 16, (1), pp. 65-73, 1995

Kolodner, d.: 'Improving human decision making through case-based decision aiding', *AI Magazine*, 12, (2), pp. 52-68, 1991

Kourti, T., and J. F. Macgregor: 'Process analysis, monitoring and diagnosis, using multivariate projection methods', *Chemometrics and Intelligent Laboratory Systems*, 28, pp. 3-21, 1995

Kourti, T., J. Lee and J.F. Macgregor: 'Experiences with industrial applications of projection methods for multivariate statistical process control', *Computers and Chemical Engineering*, 20, pp. 745-750, 1996

Kresta, J.V., T.E. Martin and J.F. MacGregor: Development of inferential process models using PLS', *Computers and Chemical Engineering*, 18, (7), pp. 597-611, 1994

Lewin, D.R. 'Predictive maintenance using PCA', *Control Engineering Practice*, 3, (3), pp. 415-421, 1995

Li, W., H.H. Yue, S. Valle-Cervantes, and S.J. QIN: 'Recursive PCA for adaptive process monitoring', *Journal of Process Control*, 2000, 10, pp. 471-486

Li, G. and Liu, B.: "RBFNN algorithm based on hybrid hierarchy genetic algorithm and its application". *Control Theory & Applications*. 19(4), pp. 627-630, 2002.

MacGregor, J.F., and T. KOURTI: 'Statistical process control of multivariate processes' *Control Engineering Practice*, 3, (3), pp. 403-414, 1995

MacGregor, J.F., H. YU, S.G. Munoz and J. Flores-Cerrillo: 'Data-based latent variable methods for process analysis, monitoring and control', *Computers and Chemical Engineering*, 29, pp. 1217-1223, 2005

Martin, E.B., A.J. Morris and J. Zhang: 'Process performance monitoring using multivariate statistical process control', *IEEE Proceedings Control Theory and Applications*, 143, (2), pp. 132-144, 1996

Matthews, R.: 'Data miners only strike fool's gold', *New Scientist*, 8th March, pp. 8, 1997

Michalski, R.S., I. Bratko and M. Kubat" 'Machine learning and data mining: Methods and applications', Wiley & Sons, Chichester, England, 1999

Moody, Darken C.: "Fast Learning in Networks of Locally-tuned Processing Units". *Neural Computation*, 1, pp. 281 – 294,1989.

Olaru, C., and L. Wehenkel: 'Data Mining', *IEEE Computer Applications in Power*, 12, pp. 19-25, 1999

Oliveira-Esquerre K.P., D.E. Seborg, R.E. Bruns and M. Mori: 'Application of steady-state and dynamic modelling for the prediction of the BOD of an aerated lagoon at a pulp and paper mill Part I. Linear approaches', *Chemical Engineering Journal*, 104, pp. 73-81, 2004

Otto, M., and W. Wegscheider: 'Spectrophotometric multi-component analysis applied to trace metal determinations', *Analytical Chemistry*, 57, (1), pp. 63-69, 1985

Pan, L., D. Flynn and M. Cregan: 'Statistical model for power plant performance monitoring and analysis', *Universities Power Engineering Conference*, pp. 121-126, 2008

Qin, S., H. Yue, and R. Dunia: 'Self-validating inferential sensors with application to air emission monitoring', *Industrial Engineering Chemical Research*, 36, pp. 1675-1685, 1997

Quinlan, J.R.: 'C4.5: programs for machine learning', 1993, The Morgan Kaufmann series in machine learning, Morgan Kaufmann Publishers, California

Sebzalli, Y.M., and X.Z. Wang.: 'Knowledge discovery from process operational data using PCA and fuzzy clustering', *Engineering Applications of Artificial Intelligence*, 2001, 14, (5), pp. 607-616

Smyth, B., M.T. Keane and P. Cunningham: 'Hierarchical case-based reasoning integrating case-based and decompositional problem-solving techniques for plant-control software design', *IEEE Transactions on Knowledge and Data Engineering*, 13, (5), pp. 793-812, 2001

Valle, S., W. Li and S.J. Qin: 'Selection of the number of principal components: The variance of the reconstruction error criterion with a comparison to other methods', *Industrial Engineering Chemistry Research*, 38, pp. 4389-4401, 1999

Voumvoulakis, E.M. and Hatziargyriou, N.D.: 'A Particle Swarm Optimization Method for Power System Dynamic Security Control', *IEEE Transactions on Power Systems*, pp. 1032-1041, 2010

Watson, I.: 'Case-based reasoning is a methodology not a technology', *Knowledge-Based Systems*, 12, pp. 303-308, 1999

Wang, J.R.: 'Research on web-based multi-agent system for aeroengine fault diagnosis', *The 9th International Conference on Web-Age Information Management*, pp. 195-202, 2008

Weiss, S.M., and N. Indurkhya: 'Predictive data mining, a practical guide', 1998, Morgan Kaufmann Publishers Inc., San Francisco, CA

Wise, B.M. and N.B. Gallagher: 'The process chemometrics approach to process monitoring and fault detection', *Journal of Process Control*, 6, (6), pp. 329-348, 1996

Wold, S., K. Esbensen and P. Geladi: 'Principal component analysis', *Chemometrics and Intelligent Laboratory Systems*, 1987, 2, pp. 37-52

Yang, P.: 'A case-based reasoning with feature weights derived by BP network', *Workshop on Intelligent Information Technology Application, IITA*, pp. 26-29, 2007

Yoon, S., and J.F. Macgregor: 'Fault diagnosis with multivariate statistical models. Part one: using steady state fault signatures', *Journal of Process Control*, 11, pp. 387-400, 2001

Part 3

Heat Transfer

9

Influence of Heat Transfer on Gas Turbine Performance

Diango A., Périlhon C., Danho E. and Descombes G.

Laboratoire de génie des procédés pour l'environnement,
l'énergie et la santé Conservatoire national des arts et métiers,
Paris
France

1. Introduction

In the current economic and environmental context dominated by the energy crisis and global warming due to the CO_2 emissions produced by industry and road transportation, there is an urgent need to optimize the operation of thermal turbomachinery in general and of gas turbines in particular. This requires exact knowledge of their typical performance.

The performance of gas turbines is usually calculated by assuming an adiabatic flow, and hence neglecting heat transfer. While this assumption is not accurate for high turbine inlet temperatures (above 800 K), it provides satisfactory results at the operating point of conventional machines because the amount of heat transferred is generally low (less than 0.5% of thermal energy available at the turbine inlet). Internal and external heat transfer are therefore neglected and their influence is not taken into account.

However, current heating needs and the decentralized production of electrical energy involve micro Combined Heat and Power (CHP) using micro-gas turbines (20-250 kW).

In aeronautics, the need for a power source with a high energy density also contributes to interest in the design of ultra-micro gas turbines.

These ultra and micro machines, which operate on the same thermodynamic principles as large gas turbines, cannot be studied with the traditional adiabatic assumption, as has been underlined by many authors such as Ribaud (2004), Moreno (2006) and Verstraete *et al.* (2007). During operation, heat is transferred from the turbine to the outside, bearing oil, casing and compressor, thus heating the compressor and leading to a drop in turbine performance. Consequently, the performances reported on the maps developed under the adiabatic assumption are no longer accurate.

This chapter presents:

• The influence of heat transfer on the performance at an adiabatic operating point of a gas turbine, and a method for determining the actual operating point knowing the amount of heat transfer.

• A study of heat loss versus the geometry scale of the volute and some conclusions concerning the limits of validity of the adiabatic assumption.

2. Relocating an adiabatic operation point subjected to heat transfer on a gas turbine map

2.1 Introduction

The performance of a turbomachine is usually represented graphically with dimensionless coordinates obtained under the assumption of adiabaticity from an existing machine.

These maps are employed by manufacturers and users to determine the overall performance in order to design a new machine or to use the same machine in different operating conditions. The results obtained are not always accurate, however, as this assumption is not valid in all circumstances. Under the influence of heat transfer, the supposedly adiabatic operating point may shift its position. The dimensionless coordinates change, making it necessary to find the actual values for a correct assessment of performance.

In order to simulate the movement of an adiabatic operating point subjected to heat transfer, we consider the single-shaft gas turbine with a simple cycle; the maps are shown in Fig. 13 (Pluviose 2005).

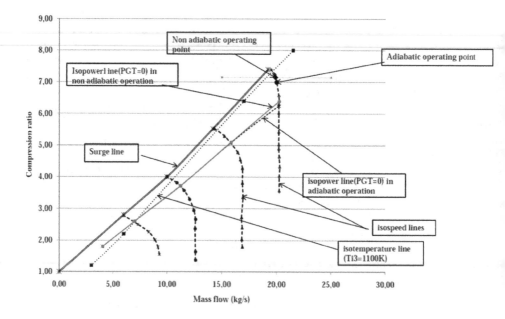

Fig. 1. Adiabatic compressor map of the gas turbine studied (Pluviose, 2005)

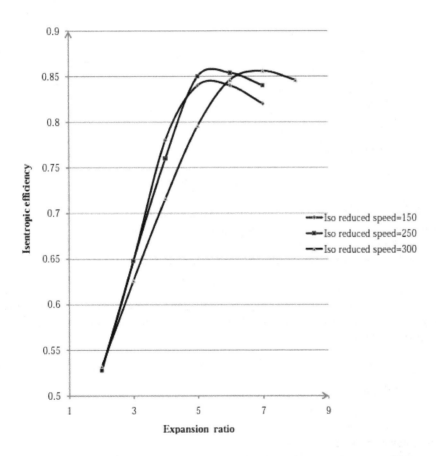

Fig. 2. Isentropic efficiency versus expansion ratio of the turbine (Pluviose, 2005)

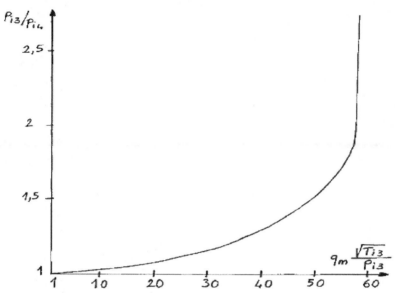

P_{i3}: turbine inlet pressure;
P_{i4}: turbine outlet pressure;
T_{i3}: turbine inlet temperature;
T_{i4}: turbine outlet temperature.

Fig. 3. Expansion ratio versus dimensionless mass flow rate (Pluviose, 2005)

2.2 Adiabatic, insulated and non insulated gas turbine versions

In its simplest form, as shown in Fig. 4, a gas turbine consists of:

- A centrifugal or axial air compressor;
- A combustion chamber in which a mixture of air and fuel is burnt;
- A centripetal or axial turbine;
- A user device (alternator, pumps, etc.).

Neglecting the kinetic and potential energy, the formulation of the first law of thermodynamics in an open system applied to turbomachinery (compressor and turbine) is written:

In transient conditions:

$$dh = \delta w + \delta q \tag{1}$$

In steady conditions:

$$\Delta h = w + q \tag{2}$$

dh: elementary variation of enthalpy;
      δw: elementary work exchanged;
      δq: elementary amount of heat exchanged with the surroundings;
      Δh: specific enthalpy variation;
      w: specific work exchanged by the fluid with the moving parts of the machine;
      q: heat exchanged by the fluid with its surroundings.

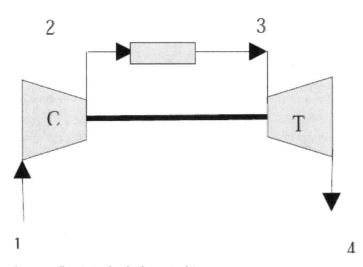

Fig. 4. Simple-cycle, open flow, single-shaft gas turbine

In conventional machines, calculations are usually done by assuming that the gas turbine is adiabatic (q = 0).

The adiabatic version of a gas turbine is one in which the heat exchanged by the fluid with surroundings in the turbomachine is exactly zero (compressor: q_{12}; turbine: q_{34}). This version cannot be obtained in practice because of the difference in temperature between the turbine inlet and the surroundings. In order to approximate this ideal configuration, experimenters introduce some thermal insulation. This leads to the concept of insulated and non-insulated gas turbines.

In an insulated gas turbine, the fluid in the turbomachine is assumed not to exchange thermal energy with the surroundings. In practice, this is achieved by insulating the machines with very low thermally conductive materials. However, because of the external insulation, internal heat exchange (in particular from the turbine to the compressor) is increased and must be taken into account.

The non insulated gas turbine is equivalent to one in which internal and external heat transfer coexist.

2.3 Characteristics of the nominal operating point of an adiabatic gas turbine (Pluviose, 2005)

The assumptions are:

- Power of mechanical losses: P_{ml} = 66 kW;
- Turbine inlet temperature: T_{i3} =973 K;
- Isentropic efficiency of the turbine: η_T = 0.85;
- Rotational speed : N=8000 rpm;
- The compressor mass flow rate: q_m: 20 kg.s^{-1};
- Inlet conditions : p_{i1} =1.013 bar et T_{i1} = 288 K ;
- Specific heat at constant air pressure: c_p=1 kJ/kg
- Specific heat at constant pressure of the burnt gas: c_p=1.13 kJ/kg

- Compression ratio : $\pi_c = 7$;
- Isentropic efficiency of the compressor : $\eta_C = 0.8$;
- Pressure drop in combustion chamber: 5%;
- Specific heat ratios : for air $\gamma = 1.4$; for burnt gas $\gamma = 1.33$;
- Turbine outlet pressure: $p_{i4} = 1.05$ bar.

The characteristics of the operating point of the adiabatic gas turbine are summarized in Table 1.

	q_m (kg.s^{-1})	π_C or π_T	N (rpm)	P_{GT} (kW)	T_{i3} (K)	Q_{cc} (kW)	η_{GT} %
Adiabatic compressor	20	7	8 000	1 526.4	973	9 431	16.2
Adiabatic turbine	20	6.42	8 000				

Table 1. Characteristics of the operating point of an adiabatic gas turbine

The energy balance at the operating point is shown in Table 2 (see the detailed calculations in the appendix).

	Energy balance	
	kW	%
Thermal power provided by fuel	9430.7	100
Gas Turbine power	1526.2	16.2
Thermal power loss at exhaust	7838.5	83.12
Mechanical losses	66	0.7
Thermal losses	0	0

Table 2. Energy balance at the operating point of the adiabatic gas turbine

2.4 Characteristics of the nominal operating point of a non-adiabatic gas turbine
As indicated in section 2.2, there are two non-adiabatic versions of the gas turbine: the insulated and the non-insulated version.

2.4.1 Influence of heat transfer on the adiabatic nominal operating point
In order to understand the influence of heat transfer on the nominal operating point, we assume that the turbine is cooled so that the heat losses account for 15% of the adiabatic work. For the non-insulated version, 60% of these losses are considered to contribute to the heating of the compressor (Rautenberg & al. 1981).

In the insulated version, it is assumed that all the heat lost by the turbine is received by the air in the compressor.

In this study, the amount of heat exchanged is assumed known. The internal work depends on the outlet temperature. In practice, during operation, the outlet temperature of the machine can be measured. But, here, we choose $T_{i2} > T_{i2ad}$ (compressor) and $T_{i4is} < T_{i4} < T_{i4ad}$ (turbine).

Table 3 and Table 4 summarize the new performances calculated for the adiabatic gas turbine used in insulated and non-insulated versions at the adiabatic operating point.

	q_m (kg^{-1})	π_C ou π_T	N (rpm)	P_{GT} (kW)	T_{i3} (K)	Q_{cc} (kW)	η_{GT} %
Heated compressor	20	7	8 000	1 128	973	8 521	13.2
Cooled Turbine	20	6.42	8 000				

Table 3. Characteristics of nominal operating point in non insulated version

	q_m (kg^{-1})	π_C ou π_T	N (rpm)	P_{GT} (kW)	T_{i3} (K)	Q_{cc} (kW)	η_{GT} %
Heated compressor	20	7	8 000	1010.38	973	7917	12.8
Cooled Turbine	20	6.42	8 000				

Table 4. Characteristics of nominal operating point in insulated version

The comparison of the results in Table 1, Table 3 and Table 4, leads to the following comments:

- Energy efficiency has dropped from 16.2 to 13.2% (Table 1 and Table 3), and from 16.2 to 12.8% (Table 1 and Table 4);
- Net power of the gas turbine has decreased from 1 526 to 1 128 kW, or by 26% (Table 1 and Table 3); from 1526 to 1010 kW, or by 34% (Table 1 and Table 4);

We can therefore conclude that if the gas turbine operates with heat transfer while maintaining the same parameters as under nominal adiabatic operation, there is a drop in performance.

This significant drop in performance makes it necessary to determine the actual operating point, taking into account heat transfer and the needs of user devices. For example, in a power plant equipped with a gas turbine, meeting the needs of the consumer requires that the power be kept constant. This involves finding the new non-adiabatic operating point which fulfills this criterion (same power at constant rotational speed).

2.4.2 Search for a new operating point able to provide the same power

Search for the new operating point of the compressor

The gas turbine operates under adiabatic or non-adiabatic conditions at 8000 rpm. For this speed, the output power is plotted versus the compression ratio in the three configurations: adiabatic, insulated and non-insulated versions (Figure 5).

For the selected power value, the new compression ratios in insulated and non-insulated operation can be deduced. Then drawing this pressure ratio on the compressor map (Fig. 6), the mass flow rate and the efficiency of this point are deduced.

Comments:

It can be seen on Figure 5 that for the same compression ratio, the net output power is low in the insulated version. The highest output power is obtained in the adiabatic version.

For the same power, the compression ratio is low in the non-insulated version. The lowest value is obtained in the adiabatic version.

Search for the new operating point of the turbine

As the rotational speed is constant and imposed, the required power can be achieved only by means of the quantity of injected fuel which has a direct influence on T_{i3} (turbine inlet temperature).

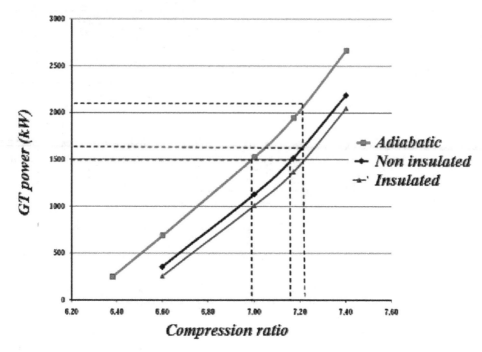

Fig. 5. Output power of the gas turbine versus compression ratio in adiabatic, non-insulated and insulated version

Due to the turbine characteristics, for a pressure ratio above that shown in Fig.3, the reduced mass flowing through the turbine is a constant which was calculated for the nominal operating point in adiabatic conditions (Pluviose, 2005). A reduced mass flow makes it possible to determine the new value of T_{i3} corresponding to the new pressure p_{i3}.

$$q_m \frac{\sqrt{T_{i3}}}{p_{i3}} = cst \qquad (3)$$

q_m: mass flow rate (kg.s^{-1});
T_{i3}: turbine inlet temperature (K);
p_{i3}: turbine inlet pressure (p$_a$).

Non-insulated gas turbine

The characteristics of the new operating points are summarized in table 5 (see calculations in the appendix).

Comparing the results of Table 1 to Table 5, it can be observed that:

- The mass flow rate has decreased. It varies from 20 to 19.8 kg^{-1}. The relative deviation is 1%;
- The compression ratio has increased from 7 to 7.17 with a relative deviation of 2.4%;
- The turbine inlet temperature has risen from 973 to 1 041 K. The maximum is 1100 K;
- The energy efficiency has decreased to 16.2 à 15.6% (the relative deviation is 3.7%).

	q_m (kg.s⁻¹)	π_C ou π_T	N (rpm)	P_{GT} (kW)	T_{i3} (K)	Q_{cc} (kW)	η_{GT} %
Heated compressor	19.8	7.17	8 000	1526.2	1040.6	9764	15.6
Cooled turbine	19.8	6.57	8 000				

Table 5. Characteristics of the new operating point *in the non-insulated version*

Insulated gas turbine

In order to simplify calculations, we consider that all the heat lost by the turbine is fully received by the compressor

	q_m (kg.s⁻¹)	π_C ou π_T	N (rpm)	P_{GT} (kW)	T_{i3} (K)	Q_{cc} (kW)	η_{GT} %
Heated compressor	19.5	7.22	8 000	1526.2	1088	10253	14.9
Cooled turbine	19.5	6.62	8 000				

Table 6. Characteristics of the new operating point in the *insulated version*

When the results of tables 1 and 6 are compared, it can be seen that:
- The mass flow rate has decreased from 20 to 19.5 kg.s⁻¹. The relative deviation is 2.5%;
- The compression ratio has increased from 7 to 7.22. The relative increase is 3.14%;
- The turbine inlet temperature has increased from 973 to 1 088 K. The limit is   1 100 K;
- The energy efficiency has dropped from 16.2 to 14.9% (the relative deviation is 8.02%).

Overall in the two operating configurations, the operating area on the compressor map has slightly narrowed.

However, the temperature increase can be a problem, as this value has a direct influence on the turbine life span.

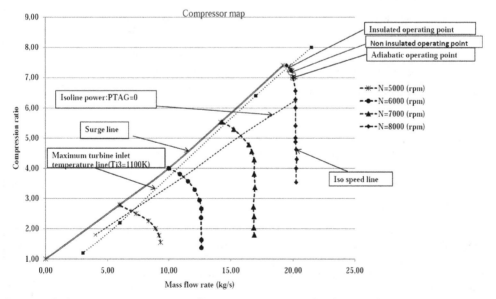

Fig. 6. Adiabatic compressor map with operating points in the three configurations

2.4.3 Energy balance of new operating points

Non-insulated gas turbine

	Energy balance	
	kW	%
Calorific power provided by the fuel	9764	100
GT power	1526.2	15.6
Exhaust power	7129.6	73.0
Mechanical losses	66	0.7
Thermal losses	1042	10.7

Table 7. Energy balance of the new operating point (non-insulated gas turbine)

Insulated gas turbine

	Energy balance	
	kW	%
Calorific power provided by the fuel	10253	100
GT power	1526.2	14.9
Exhaust power	7618.8	74.3
Mechanical losses	66	0.6
Thermal losses	1042	10.2

Table 8. Energy balance of the new operating point (insulated gas turbine)

Comparing tables 7 and 8, we can see that at iso speed and iso net power produced, the efficiency of the gas turbine is better in the non-insulated version.

2.5 Comparison with experimental results

The analysis and the results presented above for the nominal operating point were extended to the other points of the working area.

Figure 8 shows the experimental results obtained by Moreno (2006) on a small gas turbine (75 kW).

The tests were carried out in two versions: an insulated version at 39 000 rpm and a non-insulated one at 40 000 rpm. It may be noted that the speeds are not identical because of the practical difficulties of measurement in testing. But the relative difference of 2.5% between these two speeds can be considered negligible.

Figure 9 shows that, as in the case of our study, iso-speed, iso net power produced by the gas turbine, and energy efficiency are better in the insulated than in the non-insulated version.

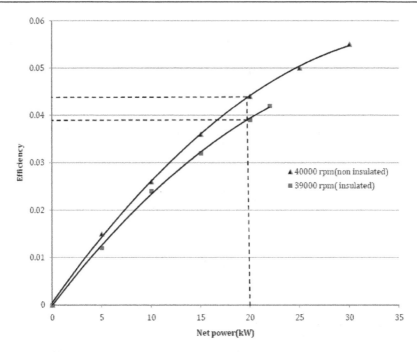

Fig. 7. Energy efficiency versus net power produced (Moreno, 2006)

This study not only confirms the decrease in performance due to heat losses, but also that this drop in performance is proportionally greater with the internal heat transfer. Compared with the gas turbine studied here (1500 kW), it can be seen that the energy efficiency of the gas turbine used by Moreno (75 kW) is very low: 8% vs 16% at nominal power. This can be attributed to the size of the machine: it is a small machine with a nominal power about approximately twenty times smaller. Heat losses could be the cause of the drop in performance.

3. Heat transfer and geometric scale of gas turbines

As already mentioned in the introduction, the results of the performance calculations carried out in conventional turbomachines remain satisfactory at the full load operating point. In addition, the literature indicates that the impact of heat transfer on the performance of small turbomachines is negative. In these circumstances, it is important to know the characteristic size of the machines in which the assumption of adiabaticity is no longer valid.

This study of heat transfer limited only to the volute of machines studied is conducted in similar operating conditions. It therefore calls on the notion of similarity.

3.1 The similarity of turbomachines: a summary

Similarity makes it possible, when a physical phenomenon for given operating conditions is known, to predict the same phenomenon for other conditions through laws involving dependent and independent dimensionless variables. Similarity generally focuses on two aspects: the geometric aspect that is relative to a family of geometrically similar machines, and the functional aspect that deals with a family of machines with similar operation. These two aspects are simultaneously taken into account.

For adiabatic machines, the dimensionless independent variables used to characterize the similar operating points are (Pluviose, 2005):

- dimensionless mass flow rate
- dimensionless speed.

For adiabatic machines, the dimensionless independent variables used to characterize the similar operating points are (Pluviose, 2005):

- Dimensionless mass flow rate:

$$\frac{q_m \sqrt{r T_{i1}}}{p_{i1} R^2}$$

- Dimensionless rotational speed:

$$\frac{NR}{\sqrt{r T_{i1}}}$$

q_m : mass flow rate (kg.s^{-1});
T_{i1} : turbomachine inlet temperature(K);
p_{i3} : turbomachine inlet pressure (p_a);
R : external radius of the rotor a (m);
r : specific perfect gas constant (J.kg^{-1}).

A study of similarity in non-adiabatic turbomachines operating with compressible fluid, conducted by Diango (2010) led to the generalization of Rateau's theorem. The author shows that these two dimensionless variables are also valid when operating with heat transfer.

From the foregoing and for a judicious comparison of heat exchange in different volutes, it is generally assumed that the fluid flows are similar. This leads to the following assumptions:

- Inlet parameters are the same (pressure and temperature);
- Reynolds numbers are equal;
- The inlet dimensionless velocities are identical;
- The mass flow rate is the same in the volutes.

In the heat transfer equations, only the mass flow rate and the inlet conditions are involved. For two machines a and b, the first and last assumptions imply:

$$\frac{q_{ma} \sqrt{r T_{i1}}}{R_a^2 p_{i1}} = \frac{q_{mb} \sqrt{r T_{i1}}}{R_b^2 p_{i1}} \Leftrightarrow q_{mb} = \frac{q_{ma} R_b^2}{R_a^2}$$

$$q_{mb} = \frac{q_{ma} R_b^2}{R_a^2} \qquad (4)$$

q_{ma}: mass flow rate of machine a (kg.s^{-1});
q_{mb}: mass flow rate of machine b (kg.s^{-1});
p_{i1}: inlet pressure (p_a);
T_{i1}: inlet temperature (K);

R_a: external radius of the rotor of machine a (m);
R_b: external radius of the rotor of machine b (m);
r: specific perfect gas constant (J.kg⁻¹).

Due to the complex geometry of the casing (volute) of turbomachines and the difficulties of calculating heat transfer coefficients, a numerical approach has been adopted.

3.2 Mathematical modeling of heat transfer in the gas turbine volute

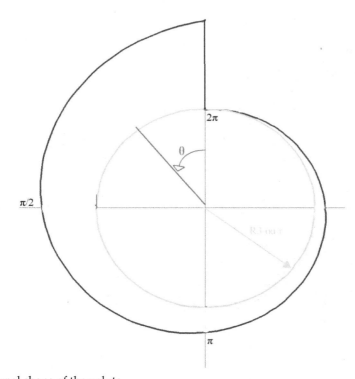

Fig. 8. External shape of the volute

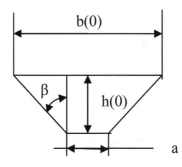

h (0): Inlet height (m)

Fig. 9. Inlet section of the volute (θ = 0)

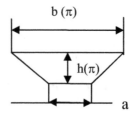

$h(\pi)$: height at $\theta = \pi$

Fig. 10. Section at $\theta = \pi$

The outer shape is a logarithmic spiral.

3.2.1 Energy balance on a mesh volume

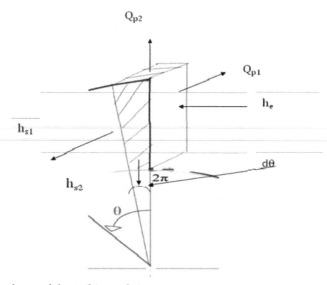

Fig. 11. Mesh volume of the turbine volute

The first law of thermodynamics can be written in any time by the following equation:

$$\frac{dU}{dt} = \sum_i \dot{q}_i + \sum_j \dot{m}_j h_j^* + \sum_k \dot{w}_k \tag{5}$$

$$h_j^* = \left(h_j + \frac{v_j^2}{2} + gz_j \right) \tag{6}$$

$\dot{q}_i$: Heat exchanged by the fluid with surroundings (W);

h_j^* : Total specific enthalpy of component j in the mesh volume (J.kg^{-1});

h_j : Specific enthalpy of component j in the mesh volume (J.kg^{-1});

$\dot{w}_k$: Mechanical power exchanged by surroundings (W)

$\dot{m}_j$: Flow of material exchanged by the component j with surroundings (W);

$\dfrac{v_j^2}{2}$: Kinetic energy of component j;

gz_j : Potential energy of component j;

In the volute, the fluid does not exchange mechanical energy with the surroundings, so:

$$\sum_k \dot{w}_k = 0$$

Variations in kinetic and potential energy are generally negligible compared to the enthalpy. Equation (5) becomes:

$$\frac{dU}{dt} = \sum_i \dot{q}_i + \sum_j \dot{m}_j h_j \tag{7}$$

In steady state, equation (7) becomes:

$$q_{m\,in} h_{in} - q_{m\,out1} h_{out1} - q_{m\,out2} h_{out2} - \dot{q}_{w1} - \dot{q}_{w2} = 0 \tag{8}$$

$\dot{q}_{w1,2}$: Heat exchanged through the walls;

$h_{in1,2}$: Total specific enthalpy at inlet of the mesh volume (J.kg^{-1});

$h_{out1,2}$: Total specific enthalpy at outlet of control volume (J.kg^{-1});

$\dot{w}_k$: Mechanical power exchanged by the fluid with surroundings (W);

$$\dot{q}_p = \dot{q}_{p1} + \dot{q}_{p2}$$

It is assumed in a first approximation that the specific heat of the burnt gas at constant pressure (cpf) does not vary in the mesh volume and that the inlet temperature of gas in the upstream guide is identical to the outlet of the volume vi. Knowing that:

$$q_{m_{in}} = q_{m_{out1}} + q_{m_{out2}}$$

Noted henceforth as:

$$q_{m_{in}} = q_{m_{i-1}}$$

The expression (8) becomes :

$$q_{m_{i-1}} c_p T_{i-1} = q_{m_{i-1}} c_p T_i + \dot{q}_p \tag{9}$$

$$\dot{q}_p = h_{eq} S \left(T_{burnt\,gas} - T_a \right) \tag{10}$$

S : Heat exchange surface (m²);
h_{eq} : Equivalent convective heat transfer coefficient (W.m⁻².K⁻¹);
$T_{burnt\,gas}$: Average temperature of the burnt gas in the control volume (K);
T_a : Average ambient temperature (K)

3.2.2 Heat exchange surface
The mesh volume is considered as a tube in which fluid flows. The elementary exchange surface is the area of the contour of the tube, not including the passage through the upstream guide. It is equal to the perimeter multiplied by the elementary length of the spiral.
The equation of the outer profile of the logarithmic spiral volute ranging from 0 to 2π is given by equation (11), following Moreno (2006):

$$R(\theta) = ae^{-b\theta} + c \tag{11}$$

$R(0)$, $R(\pi)$ and $R(2\pi)$: external radius of the volute at $\theta = 0$; π and 2π (m).
Assuming $R(0)$, $R(\pi)$ and $R(2\pi)$ are known, we can write the following relationships, from Moreno (2006), which describe the outer shape of the volute:

$$c = \frac{R(2\pi)R(0) - R^2(\pi)}{R(2\pi) + R(0) - 2R(\pi)} \tag{12}$$

$$b = -\frac{1}{2\pi} \ln \left[\frac{R(2\pi) - c}{a} \right] \tag{13}$$

$$a = R(0) - c \tag{14}$$

a and b: constant real numbers (m)
The arc length of the spiral at the position θ is given by equations (15) and (16), from Berger & Gostiaux (1992):

$$L(\theta) = \int_a^b \sqrt{R'(\theta)^2 + R(\theta)^2} \, d\theta \tag{15}$$

$$L(\theta) = \int_u^v \sqrt{a^2 b^2 e^{-2b\theta} + \left(ae^{-b\theta} + c \right)^2} \, d\theta \tag{16}$$

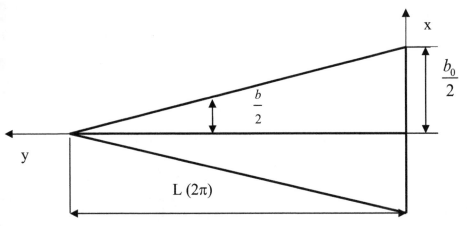

Fig. 12. The width of the upper volute versus the unrolled spiral

The flow section has a trapezoidal shape whose dimensions must be calculated in order to estimate the perimeter.

To determine the large base b (θi), of the trapezium, the equation used is:

$$b(\theta)=b_0\left[1-\frac{L(\theta)}{L(2\pi)}\right]$$ (17)

The angle generated by the height and the lateral side (Figure 9, Figure 10) is expressed by:

$$tg\beta = \frac{b(\theta)-l_{tn}}{h(\theta)}$$ (18)

l_{tn}: turbine nozzle width.

The length of the lateral side (L_l) is:

$$L_l(\theta) = \frac{h(\theta)}{\cos\beta}$$ (19)

Hence the perimeter is:

$$P(\theta) = 2L_l(\theta) + b(\theta)$$ (20)

The heat exchange surface of the mesh volume is then:

$$dA = PdL = \left[2L_l(\theta) + b(\theta)\right]dL$$

$$A = \left[2L_l(\theta) + b(\theta)\right]\int_u^v \sqrt{a^2b^2e^{-2b\theta} + \left(ae^{-b\theta} + c\right)^2}\,d\theta$$ (21)

3.2.3 Heat exchange coefficient

Forced convection inside the volute

The heat transfer coefficient is given by equation (22)

$$h_{fc}(v) = \frac{Nu(v_i)\lambda_{burnt\,gas}}{D_h(v_i)} \tag{22}$$

hfc: coefficient of forced convection ;
λburnt gas: conductivity of burnt gas;
Dh: hydraulic diameter;
vi: mesh volume.
The forced convection model uses the simplified correlation of Sieder & Tate (Kreith 1967) to calculate the Nusselt number (Nu) from the Prandtl (Pr) and Reynolds (Re) numbers:

$$Nu = 0{,}023 Re^{0{,}8} Pr^{0{,}3} \tag{23}$$

$$Pr = \frac{\mu_{burnt\,gas} C_{p\,burnt\,gas}}{\lambda_{burnt\,gas}} \tag{24}$$

$$Re(v_i) = \frac{q_m D_h(v_i)}{\mu_{burnt\,gas} S(v_i)} \tag{25}$$

μburnt gas: dynamic viscosity of burnt gas;

Free convection outside the volute

The Nusselt number is given by the Morgan correlation (Padet 2005) from the Rayleigh number (Ra).

$$\overline{Nu} = A R_a^n \tag{26}$$

$$Ra = \frac{g\beta D_h^{3}(T_w - T_a)}{\alpha v_a} \tag{27}$$

T_W: wall temperature;
v_a : Kinematic viscosity of the air;
T_a: air temperature.
The coefficient of external heat exchange is the combination of a coefficient by natural convection and exchange by radiation defined by equation 28.

$$h_{amb} = h_{nc} + h_{rad} \tag{28}$$

hamb: coefficient of heat exchange with ambient air;
hnc: coefficient of heat exchange by natural convection;
hrad: coefficient of heat exchange by radiation.

The coefficient of heat exchange by radiation is given by equation (29):

$$h_{rad}(v_i) = \frac{\varepsilon\sigma\left(T_w^4 - T_a^4\right)}{T_w - T_a}$$ (29)

The equation of the model of mass flow in the volume is (30):

$$q_{m\theta} = q_{m0}\left(1 - \frac{\theta}{2\pi}\right)$$ (30)

$q_{m\theta}$: the mass flow at position θ;
q_{m0}: the mass flow at position 0.

3.3 Numerical results of modeling

Calculations are started from a nominal operating point of a turbine of a turbocharger whose dimensions are known (see Table 9).
Geometric similarity requires that the linear dimensions of the similar volutes be multiplied by the same factor. Four other turbines have therefore been considered. The main dimensions and corresponding mass flow are recorded in table 10, which also shows the numerical results of the modeling.

T_{i3} (K)	q_m (kg.s^{-1})	T_w (K)	N (rpm)
761.3	0.848	760	110 000

Table 9. Nominal point of the reference turbine

$$Q_{wdim} = \frac{Q_w}{q_m C_p T_{i3}}$$ (31)

Q_w: Thermal power lost through the walls of the volute;
Q_{wdim}: dimensionless thermal power lost through the walls of the volute.

R(0) (m)	R(π) (m)	R(2π) (m)	b (m)	q_m (kg.s^{-1})	$\frac{S}{V}$ (m^{-1})	Q_w (W)	Q_{wdim} x10^{-4}
0.03104	0.0266	0.020325	0.017	0.0212	463	202	106
0.04656	0.0399	0.03049	0.0255	0.0477	308	273	64
0.06208	0.05321	0.04065	0.034	0.0848	231	341	45
0.0776	0.06651	0.05081	0.0425	0.1325	185	408	34
0.09312	0.07981	0.060975	0.05099	0.1908	154	474	28

Table 10. Heat loss through the volutes of different sizes

Figures 13 and 14 give the modeling results. Figure 13 shows the evolution of the heat exchange surface versus the inlet radius. The greater the volute, the smaller the surface to volume ratio. Small turbomachines therefore have a higher surface to volume ratio.

The necessity of taking into account heat transfer in small turbomachines is largely confirmed by Figure 14: the heat losses in the volute are relatively greater.

In this study, when the inlet radius is halved, the surface to volume ratio doubles and the heat losses are multiplied by about 2.5

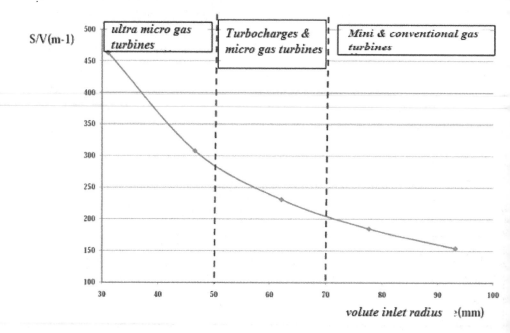

Fig. 13. Ratio of heat exchange surface (S) and the volume (V) of the volute versus the inlet radius

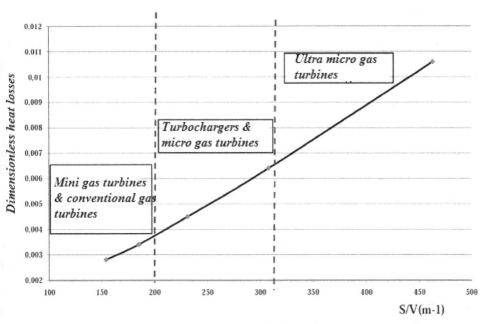

Fig. 14. Heat transfer in the volute versus the size of the machine

4. Conclusion

Internal and external heat transfer induces a drop in the performance of gas turbines. This study shows that the performance of small turbomachines evaluated with the assumption of adiabaticity is not accurate.

For a given operating point, the mass flow and the compression ratio recorded on the maps and the calculated performance do not correspond to the actual characteristics when the machine operates with heat transfer.

The assumption that heat losses represent 15% of the work of adiabatic turbines, of which 60% is received by the compressor (non-insulated), leads to overestimating the power by 35% and the energy efficiency by 23%

Insulation of the turbine, although it seems to be a solution to maintain the operating characteristics of adiabatic turbines, leads in fact to increasing the drop in performance.

For the insulated version, the net power is overestimated by 51% and efficiency by 26.6%. In the absence of an adiabatic gas turbine (ideal machine), which provides the best performance, we must avoid insulating the turbine in order not to decrease performance still further.

To maintain the level of performance, and in particular the net power produced by the gas turbines, despite heat transfer, adjustments are needed. They consist mainly in increasing the fuel flow, resulting in an increase in the turbine inlet temperature. In the case of our study, the fuel flow increase is 3.5% in the non-insulated version and 8.5% in the insulated version. The turbine inlet temperature increase is 6.4% in the insulated version and 11.8% in the non-insulated version.

Finally, this study confirms that the assumption of adiabaticity is not valid in turbochargers, micro and ultra-micro gas turbines. Compared to the available thermal energy at the turbine inlet, heat losses increase with the surface to volume ratio which decreases in small-sized machines. The quality of operation of small turbomachinery cannot be characterized with isentropic efficiency which has no physical meaning because of the relative importance of heat transfer.

The proposal of a new performance indicator and the development of new maps available for any type of thermal turbomachines will therefore be the subject of our forthcoming investigations.

5. Acknowledgment

The authors would like to acknowledge the French Cooperation EGIDE for funding this study.

6. Appendix: Energy balance calculations

1. Adiabatic gas turbine:

Data: (see page 6)
Power of the gas turbine: P_{GT}
Compressor power : Pc

$$P_C = q_m c_p \left(T_{i2} - T_{i1}\right) \; ; \; T_{i2} = T_{i1} + \left[\left(\frac{p_{i2}}{p_{i1}}\right)^{\frac{\gamma-1}{\gamma}} - 1\right]\frac{T_{i1}}{\eta_C} = 288 + \left[(7)^{\frac{0,4}{1,4}} - 1\right]\frac{288}{0,8} = 555.71\,K$$

$$P_C = q_m c_p \left(T_{i2} - T_{i1}\right) = 20 \times 1 \left(555.71 - 288\right) = 5354.2\,kW$$

Turbine power P_T:

$$P_T = q_m c_p \left(T_{i4} - T_{i3}\right) \; ;$$

$$T_{i4} = T_{i3} + \left[\left(\frac{p_{i4}}{p_{i3}}\right)^{\frac{\gamma-1}{\gamma}} - 1\right]T_{i3} \times \eta_T = 973 + \left[\left(\frac{1}{6.42}\right)^{\frac{0.33}{1.33}} - 1\right] \times 973 \times 0.85 = 665.64\,K$$

$$P_T = \left|q_m c_p \left(T_{i4} - T_{i3}\right)\right| = \left|20 \times 1.13 \times \left(665.64 - 973\right)\right| = 6946.4\,kW$$

$$P_{GT} = P_T - P_C - P_{ml} = 6946.4 - 5354.2 - 66 = 1526.2\,kW$$

Thermal power supplied by combustion chamber: Q_{cc}
The fuel flow is neglected

$$Q_{cc} = q_m c_p \left(T_{i3} - T_{i2}\right) = 20 \times 1.13 \times \left(973 - 555.71\right) = 9430.7\,kW$$

Thermal power lost in the exhaust gas: Q_{exh}

$$Q_{exh}=Q_{CC}-P_{ml}-P_{GT}=9430.7-66-1526.2=7838.5\,kW$$

2. Non insulated gas turbine:

Data: π_c = 7.17 (Figure 6); q_m = 19.8 kg.s⁻¹ (From the adiabatic compressor map). T_{i2} = 604.20 K; Q_{12} = 625.2 kW (thermal power received by the compressor).

Power of the gas turbine: P_{GT}

Thermal power received by the compressor: Q_{12}

$$Q_{12}= 0.15\times0.6\times P_{Tad}=0.15\times0.6\times 6946.4=625.2\ kW$$

P_{Tad} : adiabatic turbine power

Compressor power: Pc

$$h_2-h_1=\frac{\gamma r}{\gamma-1}(T_{i2}-T_{i1})=\frac{0.287\times1.4}{0.4}\times(604.20-288)=317.62\,kJ.kg^{-1}$$

$$\Delta H_{12}=q_m(\Delta h_{12})=317.62\times19.8=6288.9\ kW$$

$$P_C=\Delta H_{12}-Q_{12}=6288.9-625.2=5663.7\ kW$$

Turbine power P_T:

$$P_{GT}=1526.2\ kW$$
$$|P_T|=P_{GT}+P_C+P_{ml}=1526.4+5663.7+66=7256.1\,kW$$

Search for new turbine inlet temperature

The variation in the expansion ratio of the turbine versus the reduced mass flow (Figure 3) shows that when the expansion ratio is greater than two (2), the reduced mass flow remains constant (Pluviose M., 2005). This reduced flow constant calculated in adiabatic conditions enables the new turbine inlet temperature (T_{i3}) corresponding to the new pressure (p_{i3}) to be determined by the following equations.

$$\left(q_m\frac{\sqrt{T_{i3}}}{p_{i3}}\right)_{ad}=\left(q_m\frac{\sqrt{T_{i3}}}{p_{i3}}\right)_{non\,ins}=q_{m(reduced)}=\left(20\times\frac{\sqrt{973}}{6.42\times1.05\times10^5}\right)=92.55\times10^{-5}$$

ad: adiabatic
Non ins: non insulated

$$p_{i3non\,ins}=0.95\times\pi_{Cnon\,ins}\times p_{i1}=0.95\times7.17\times1.01325=6.902\,bars$$

$$T_{i3non\,ins}=\left(\frac{q_{m(reduced)}\times p_{i3non\,ins}}{q_{m\,non\,ins}}\right)^2=\left(\frac{92.55\times10^{-5}\times6.902\times10^5}{19.8}\right)^2=1040.6\,K$$

$$T_{i4isentropic}=T_{i3}\left(\frac{1}{\Pi_T}\right)^{\frac{\gamma-1}{\gamma}}=1041.6\times\left(\frac{1}{6.57}\right)^{\frac{0.33}{1.33}}=652.90\,K$$

$$T_{i4nonins}=671.23\,K$$

Thermal power supplied by the combustion chamber: Q_{cc}
The fuel flow is neglected

$$Q_{cc}=q_m c_p\left(T_{i3}-T_{i2}\right)=19.8\times1.13\times\left(1040.6-604.2\right)=9764\,kW$$

Thermal power lost in the exhaust gas

$$Q_{exh}=Q_{CC}-P_{ml}-P_{thl}-P_{GT}=9764-66-625.2/0.6-1526.4=7129.56\,kW$$

P_{thl} : power of thermal losses .

3. Insulated gas turbine:

Data: π_c = 7.22 (Figure 6); q_m = 19.5 kg.s^{-1} (from the adiabatic compressor map). T_{i2} = 622.68 K, Q_{12} = 1042 kW (thermal power received by the compressor)
Power of the gas turbine: P_{GT}
Thermal power received by the compressor: Q_{12}

$$Q_{12}=0.15\times P_{Tad}=0.15\times6946.4=1042\ kW$$

P_{Tad} : Adiabatic turbine power
Compressor power: P_C

$$h_2-h_1=\frac{\gamma r}{\gamma-1}\left(T_{i2}-T_{i1}\right)=\frac{0.287\times1.4}{0.4}\times\left(622.68-288\right)=336.19\,kJ.kg^{-1}$$

$$\Delta H_{12}=q_m\left(\Delta h_{12}\right)=317.62\times19.5=6555.6\ kW$$

$$P_C=\Delta H_{12}-Q_{12}=6555.6-1042=5513.6\ kW$$

Turbine power P_T :

$$P_{TAG}=P_T=1526.2\ kW$$
$$\left|P_T\right|=P_{TAG}+P_C+P_{ml}=1526.4+5513.6+66=7105.8\,kW$$

Search for new turbine inlet temperature

$$\left(q_m\frac{\sqrt{T_{i3}}}{p_{i3}}\right)_{ad}=\left(q_m\frac{\sqrt{T_{i3}}}{p_{i3}}\right)_{ins}=q_{m(reduced)}=\left(20\times\frac{\sqrt{973}}{6.42\times1.05\times10^5}\right)=92.55\times10^{-5}$$

ad : adiabatic

$$p_{i3ins}=0,95\times\pi_{Cins}\times p_{i1}=0.95\times7.22\times1.01325=6.95\,bars$$

$$T_{i3insulated} = \left(\frac{q_{m(reduced)} \times p_{i3insulated}}{q_{m\,insulated}} \right)^2 = \left(\frac{92.55 \times 10^{-5} \times 6.95 \times 10^5}{19.5} \right)^2 = 1088\,K$$

$$T_{i4isentropic} = T_{i3} \left(\frac{1}{\pi_T} \right)^{\frac{\gamma-1}{\gamma}} = 1088 \times \left(\frac{1}{6.62} \right)^{\frac{0.33}{1.33}} = 680.70\,K$$

$$T_{i4} = 690\,K$$

Thermal power supplied by the combustion chamber: Q_{CC}
The fuel flow is neglected

$$Q_{CC} = q_m c_p \left(T_{i3} - T_{i2} \right) = 19.5 \times 1.13 \times \left(1088 - 622.68 \right) = 10253\,kW$$

Thermal power lost in the exhaust Q_{exh}

$$Q_{ech} = Q_{CC} - P_{ml} - P_{thl} - P_{TAG} = 10253 - 66 - 1042 - 1526.2 = 7618.8\,kW$$

7. References

Berger, M., Gostiaux, B., 1992, Géométrie différentielle: variétés, courbes et surfaces *France Presses universitaires de Paris*, ISBN : 2-13-044708-2.

Cormerais, M. 2007, Caractérisation expérimentale et modélisation des transferts thermiques au sein d'un turbocompresseur d'automobile, *Thèse de doctorat de l'école centrale de NANTES*, pp. 1-243.

Diango, A., 2010, Influence des pertes thermiques sur les performances des turbomachines. *Thèse de doctorat du Conservatoire national des arts et métiers*, Paris, pp. 1-244.

Kreith, F., 1967, *Principles of heat transfer*, Masson, [trad.] Kodja Badr-El-Dine, Université d'ALEP (Syrie), Colorado, International textbook Company Scranton, Pennsylvania, 1967. pp. 1-654.

Moreno, N., 2006, Modélisation des échanges thermiques dans une turbine radiale, *Thèse de doctorat de l'École nationale supérieure d'arts et métiers*, pp. 158.

Padet, J., 2005, Convection thermique et massique, *Techniques de l'ingénieur*, BE 8206.

Pluviose, M., 2005. Conversion d'énergie par turbomachines, *Ellipses, pp. 1.277.ISBN 2-7298-2320-4*.

Pluviose, M., 2002, Machines à fluides, *Ellipses, pp. 1-276*, ISBN 2-7298-1175-3.

Pluviose, M., 2005, Similitude des turbomachines à fluide compressible, *Techniques de l'ingénieur*, BM 468007.2005.

Pluviose, M. & Perilhon, C.(2002). Mécanismes de conversion de l'énergie, *echniques de l'ingénieur*. BM 4281, 10-2002..

Pluviose , M., Perilhon, C., 2002, Bilan énergétique et applications, *Techniques de l'ingénieur*, BM 4283, 04.2003.

Rautenberg & Al., 1981, Influence of heat transfer between turbine and compressor on the performance of small turbocharger, *International Gas Turbine Congress, Tokyo*, Asme paper, 1981.

Ribaud, Y., 2004, Overall Thermodynamics Model of an Ultra Micro turbine, *Journal of Thermal Science*. 2004, Vol. 13, 4, pp. 297-301.

Sacadura, J. F., 1993, Initiation aux transferts thermiques, *Lavoisier Tec & Doc*, Vol. 4ème tirage 1993, pp. 1-439, ISBN. 2-85206-618-1.

Verstraete, T. & al., 2007, Numerical Study of the Heat Transfer in Micro Gas Turbines, *Journal of Turbomachinery*. ASME, Octobre 2007, Vol.129, DOI: 10.1115/1.2720874, pp 835-841.

10

Jet Impingement Cooling in Gas Turbines for Improving Thermal Efficiency and Power Density

Luai M. Al-Hadhrami, S.M. Shaahid and Ali A. Al-Mubarak
Associate Professor Center for Engineering Research, Research Institute
King Fahd University of Petroleum and Minerals
Saudi Arabia

1. Introduction

The gas turbine is an engine which produces a great amount of energy depending upon its size and weight. Gas turbines are used for aircraft propulsion and land based power generation. Thermal efficiency and power output (power density) of gas turbines increase with increasing turbine rotor inlet temperatures (RIT). Today there are gas turbines, which run on natural gas, diesel fuel, naphtha, methane, crude, low-Btu gases, vaporized fuel oils, and biomass gases. The last 20 years has seen a large growth in gas turbine technology which is mainly due to growth of materials technology, new coatings, and new cooling schemes. In a simple gas turbine cycle (Figure 1), low pressure air is drawn into a compressor (state 1) where it is compressed to a higher pressure (state 2). Fuel is added to the compressed air and the mixture is burnt in a combustion chamber. The resulting hot products enter the turbine (state 3) and expand to state 4 and the air exhausts. Most of the work produced in the turbine is used to run the compressor and the rest is used to run auxiliary equipment and to produce power. Figure 2 shows schematic of cross section of a small gas turbine.

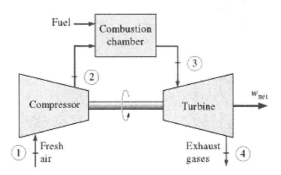

Fig. 1. Schematic of open Gas turbine cycle.

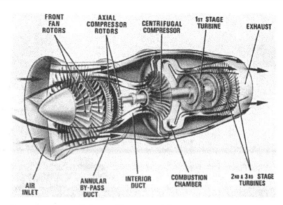

Fig. 2. A schematic of a cutaway of a small gas turbine.

2. Concept and the need for turbine blade cooling

The gas turbine engines operate at high temperatures (1200-1600 °C) to improve thermal efficiency and power output. As the turbine inlet temperature increases, the heat transferred to the turbine blades increases. The above operating temperatures are far above the permissible metal temperatures. Therefore, there is a need to cool the turbine blades for safe operation. The blades are cooled by extracted air from the compressor of the engine. Gas turbine blades are cooled internally and externally. Internal cooling is achieved by passing the coolant through several enhanced serpentine passages inside the blades and extracting the heat from outside the blades. Both jet impingement cooling and pin fin cooling are used as a method of internal cooling. External cooling is also called film cooling. Figure 3 and 4 show different types of turbine blade cooling. The cooling system must be designed to ensure that the maximum blade surface temperatures during operation are compatible with the maximum blade thermal stress.

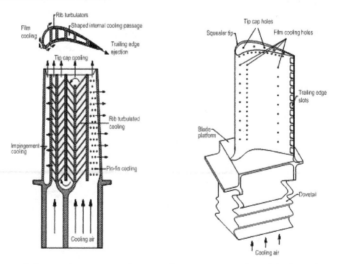

Fig. 3. Schematic of the modern gas turbine with common cooling blade techniques.

3. Typical turbine cooling system

The cooling air is bled from the compressor and is directed to the stator, the rotor, and other parts of the turbine rotor and casing to provide adequate cooling. The effect of coolant on the aerodynamics depends on the type of cooling involved. An example of a typical cooling system is shown in Figure 4.

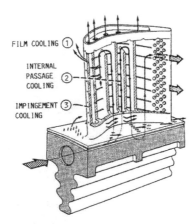

Fig. 4. Typical cooled aircraft gas turbine blade of three dimensions.

4. Jet impingement cooling

Jet impinging on the inner surfaces of the airfoil through tiny holes in the impingement insert is a common, highly efficient cooling technique for first-stage vanes. Impingement cooling is very effective because the cooling air can be delivered to impinge on the hot region. Jet impingement cooling can be used only in the leading-edge of the rotor blade, due to structure constraints on the rotor blade under high speed rotation and high centrifugal loads. Schematic of the impingement jet of the leading edge gas turbine blade is shown in Figure 5.

Several arrangements are possible with cooling jets and different aspects need to be considered before optimizing an efficient heat transfer design. Some of the research studies have focused on the effects of jet-hole size and distribution, cooling channel cross-section, and the target surface shape on the heat transfer coefficient distribution.

Wide range of parameters affect the heat transfer distribution, like impinging jet Re, jet size, target surface geometry, spacing of the target surface from the jet orifices, orifice-jet plate configuration, outflow orientation, etc. Literature indicates that many of these parameters have been studied in appreciable depth [1-20].

Chupp et al. [1] studied the heat transfer characteristics for the jet impingement cooling of the leading edge region of a gas turbine blade. Flourscheutz et al. [2] investigated the heat transfer characteristics of jet array impingement with the effect of initial crossflow. Metzger and Bunker [3] and Flourscheutz et al [4] used the liquid crystal technique to study the local heat transfer coefficients. The authors observed that the jet Nusselt number depends mainly on the jet Re.

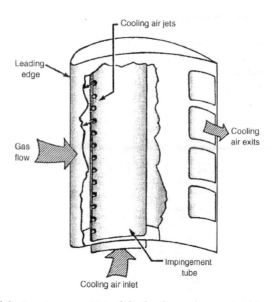

Fig. 5. Schematic of the impingement jet of the leading edge gas turbine blade.

Dong et al [5] determined experimentally the heat transfer characteristics of a premixed butane/air flame jet impinging upwards on an inclined flat plate at different angles of inclination and fixed Reynolds number (Re = 2500) and a plate to nozzle distance of 5d. It was found that decrease in angle of incidence reduced the average heat transfer. Rasipuram and Nasr [6] studied air jet issuing out of defroster's nozzles and impinging on inclined windshield of a vehicle. The overall heat transfer coefficient of the inclined surface for the configuration with one rectangular opening was found to be 16% more than that for the configuration with two rectangular openings. Beitelmal et al [7] investigated the effect of inclination of an imping air jet on heat transfer characteristics. They found that maximum heat transfer shifts towards the uphill side of the plate and the maximum Nusselt number (Nu) decreases as the inclination angle decreases. Roy and Patel [8] studied the effect of jet angle impingement on local Nu and nozzle to target plane spacing at different Re. They found that heat transfer is the maximum through the shear layer formed near the jet attachment stagnation region. Ekkad et al [9] studied the effect of imping jet angle ±45 on target surface by using transient liquid crystal technique for single Re = 1.28×10⁴. It has been noted that the orthogonal jets provide higher Nu as compared to angled jets.

Tawfek [10] investigated the effect of jet inclination on the local heat transfer under an obliquely impinging round air jet striking on circular cylinder. Their results indicated that with increase in inclination, the upstream side of heat transfer profile dropped more rapidly than the downstream side. Seyedein et.al. [11] performed numerical simulation of two dimensional flow fields and presented the heat transfer due to laminar heated multiple slot jets discharging normally into a converting confined channel by using finite difference method with different Re(600-1000) and angle of inclination (0-20°). Yang and Shyu [12] presented numerical predictions of heat transfer characteristics of multiple impinging slot jets with an inclined confinement surface for different angles of inclination and different Re. Yan and Saniei [13] dealt with measurement of heat transfer coefficient of an impinging

circular air jet to a flat plate for different oblique angles (45-90°) and different Re (10000 & 23000) by using transient liquid crystal technique.
Hwang et. al [14] studied the heat transfer in leading edge triangular duct by an array of wall jets with different Re (3000-12600) and jet spacing s/d (1.5-6) by using transient liquid crystal technique on both principal walls forming the apex. Results show that an increase in Re increases the Nu on both walls. Ramiraz et al [15] investigated the convective heat transfer of an inclined rectangular plate with blunt edge at various Re (5600-38500) and angle of inclination (60-70°). The heat transfer distribution over a finite rectangular plate was found to be very much dependent on the orientation of the plate. Stevens and Webb [16] examined the effect of jet inclination on local heat transfer under an obliquely impinging round free liquid jet striking at different Re, angle of inclination, and nozzle sizes. It was found that the point of maximum heat transfer along the x-axis gets shifted upstream. Hwang and Cheng [17] performed an experimental study to measure local heat transfer coefficients in leading edge using transient liquid crystal technique. Three right triangular ducts of the same altitude and different apex angles (30°, 45° & 60°) were tested for various jet Re (3000 ≤Re≤12000) and different jet spacing (s/d=3 and 6). Hwang and Cheng [18] measured experimentally local heat transfer coefficients on two principal walls of triangular duct with swirl motioned airflow induced by multiple tangential jets from the side entry of the duct by using transient liquid crystal technique. Hwang and Chang [19] measured heat transfer coefficients on two walls by using transient liquid crystal technique in triangular duct cooled by multiple tangential jets. The results show that an increase in Re, increases heat transfer of both walls. Hwang and Cheng [20] studied heat transfer characteristics in a triangular duct cooled by an array of side-entry tangential jets.
It is evident from the published literature that no study has been conducted to show the effect of different orifice-jet plate configurations on feed channel aspect ratio with different jet Re, for a given outflow orientation (diameter of jet, d = 0.5 cm) on heat transfer in a channel with inclined heated target surface. Therefore, the following sections include investigation of the above effects by conducting the experimental work. Specifically, the work includes the effect of three orifice-jet plate configurations (centered holes, staggered holes, tangential holes) and three feed channel aspect ratios (H/d=5, 7, and 9) on the heat transfer characteristics for a given outflow orientation (with outflow passing in both the directions) and for a different Reynolds number with inclined heated target surface. The motivation behind this work is that the channel of turbine blade internal cooling circuit at the leading edge is inclined.

4.1 Description of the experimental set-up

The schematic of the experimental set-up is depicted in Figure 6. The test rig used to study the heat transfer characteristics has been constructed using Plexi-glass. The test section consists of two channels, impingement (10) and the feed channel (9). Air enters the test section in the feed channel and is directed onto the heated copper plates in the impingement channel to study the heat transfer characteristics. The target plates (11), made of copper, were heated using a constant flux heater. The other side of the heater was insulated to get the heat transferred only in one direction. The mass flow rate of the compressed air (1) entering the test section was passed through a settling chamber (5) and was controlled with the help of valves (2). The pressure drop was measured using the pressure gauges (4). Gas flow meter (7) was used to measure the mass flow rate entering the test rig which was protected by the air filters (6) of 50μ capacity.

The average surface temperature of each copper plate was determined from the readings of two T-type thermocouples (12) installed in the holes drilled at the back surface of the plates to within 1 mm of the surface in depth. The analog signals generated by these temperature sensors were transmitted to the signal-conditioning unit where they were selectively processed. The resulting analog signals were converted into digital signals by a DAQ (13) card and recorded with application software developed in LabView.

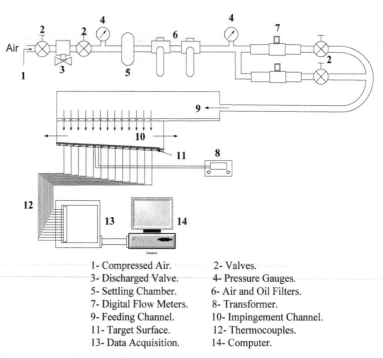

1- Compressed Air. 2- Valves.
3- Discharged Valve. 4- Pressure Gauges.
5- Settling Chamber. 6- Air and Oil Filters.
7- Digital Flow Meters. 8- Transformer.
9- Feeding Channel. 10- Impingement Channel.
11- Target Surface. 12- Thermocouples.
13- Data Acquisition. 14- Computer.

Fig. 6. Schematic of the test section

Figure 7.shows the three-dimensional sketch of the test section. It consists of two channels joined by the orifice plate, which has a single array of equally spaced (centered or staggered or tangential) orifice jets shown in Figure 8. The jet orifice plate thickness is twice the jet diameter. There are 13 jets on the orifice plate. The jet-to-jet spacing is 8 times the jet diameter and the orifice jet diameter $d = 0.5 cm$. The length of the test section is 106.5 cm. The width of the feed channel (H) was varied from 2.5 to 4.5 cm (i.e. H/d=5, 7, 9, d=0.5 cm). The impingement target surface constitutes a series of 13 copper plates, each with 4.2 × 4.1 cm in size, arranged in accordance with the orifice jets such that the impingement jet hits the geometric center of the corresponding plate (however, first and last copper plates are slightly different in sizes). All the copper plates are separated from each other by 1 mm distance to avoid the lateral heat conduction, thus dividing the target surface into segments. The thickness of the copper plate is 0.5 cm. As shown in Figure 9, the length of impingement surface L is 57.3 cm (the target surface is inclined at angle 1.5°, the width of parallel flow side "S2" is 2 cm and the width of the opposite flow side "S1" is 3.5 cm).

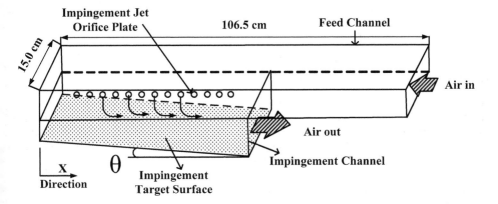

Fig. 7. Three-dimensional view of the test section

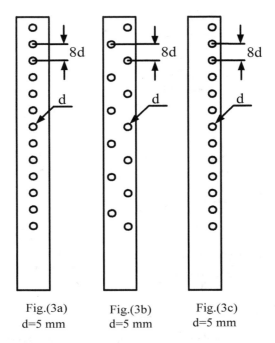

Fig. 8. Illustration of three orifice-jet configurations with single array of jets (d = 5 mm) (Fig. 8a Centered holes, Fig. 8b Staggered holes, Fig. 8c Tangential holes)

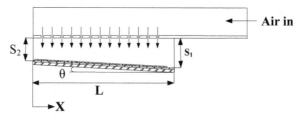

Fig. 9. Inclination angle of the target surface

Figure 10 shows the schematic of the three different outflow orientations. The upper channel is called as the feed channel and the lower channel in which the jets impinge on the target surface is called as the impingement channel. The exit of jets in three different outflow orientations from the impingement channel creates different cross-flow effects as shown in Fig. 10. However, in the present study, attention is focused on Case – 3.

Case - 1 (Outflow coincident with the entry flow),

Case - 2 (Outflow opposing to the entry flow),

Case - 3 (Outflow passes out in both the directions).

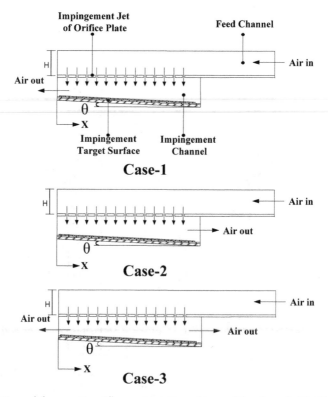

Fig. 10. Illustration of three exit outflow orientations Figure 10a, Case-1 (Outflow Coincident with the inlet flow). Figure 10b, Case-2 (Outflow Opposing to the inlet flow). Figure 10c, Case-3 (Outflow Passes out in both directions).

Figure 11 shows the details of the construction of the target surface. The copper plate is heated with a constant flux heater held between the wooden block and the copper plate by glue (to reduce contact resistance). The ends were sealed with a rubber material to avoid lateral heat losses. The wooden block was 3 cm thick to minimize the heat lost to the surroundings, so that the heat is transferred completely to copper plates only.

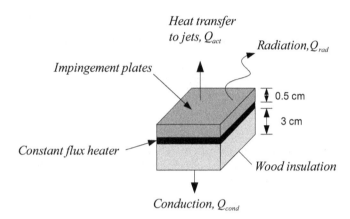

Fig. 11. Overall energy balance over a small element of the impingement plate

4.2 Experimental procedure

To start with, tests were carried out using a given orifice-jet plate (centered or staggered or tangential holes) with jet diameter $d = 0.5 cm$ for a given jet Reynolds numbers Re = 18800 (for a given H/d ratio, for outflow passing in both directions) and for a constant heat flux power input. The heated target plate was oriented at a pre-defined angle (1.5°). The mass flow rate was adjusted to the required value for the experiment to be conducted and the air was blown continuously into the test section. Heat was supplied to the copper plates with electric resistive constant flux heaters from backside to provide uniform heat flux. The temperature of the copper plates was measured by two thermocouples mounted in a groove of 2.5 mm on the back of the copper plates. Thus the temperature of a particular plate has been taken as the average of the reading of two thermocouples. The temperature of the copper plates, pressure, temperature of the air at the inlet, and the mass flow rate were continuously monitored. After the temperature of the copper plates reached the steady state condition, all the data was collected with Lab VIEW program. The Nusselt number was then calculated based upon the collected data. The same procedure was repeated for the three different orifice-jet plates described in Figures 8a, 8b & 8c and for different aspect ratios (H/d = 5, 7, 9).

4.3 Data reduction and uncertainty analysis

The collected data was subjected to uncertainty analysis. The method for performing the uncertainty analysis for the present experimental investigation has been taken from Taylor B.N. [21]. The theory for the current uncertainty analysis is summarized in the following discussion:

4.4 Jet reynolds number calculations

The average velocity used to calculate the jet Reynolds number is calculated using the following equation

$$V_{avg} = \frac{\forall}{13 \times \frac{\pi}{4} d^2} \tag{1}$$

The data reduction equation for the jet Reynolds number is taken as:

$$Re = \frac{\rho V_{avg} d}{\mu} = \frac{\rho d}{\mu} \frac{\forall}{13 \times \frac{\pi}{4} d^2} \tag{2}$$

4.5 Uncertainty in jet reynolds number

Taking into consideration only the measured values, which have significant uncertainty, the jet Reynolds number is a function of orifice jet diameter and volume flow rate and is expressed mathematically as follows:

$$Re = f(\forall, d) \tag{3}$$

Density and dynamic viscosity of air is not included in the measured variables since it has negligible error in the computation of the uncertainty in jet Reynolds number. The uncertainty in Reynolds number has been found to be about 2.2 %.

4.6 Nusselt number calculations

The total power input to all the copper plates was computed using the voltage and current, the former being measured across the heater, using the following equation:

$$Q_{total} = \frac{V^2}{R} = VI \tag{4}$$

The heat flux supplied to each copper plate was calculated using:

$$q'' = \frac{Q_{total}}{A_{total}} \tag{5}$$

The heater gives the constant heat flux for each copper plate. The heat supplied to each copper plate from the heater is calculated using the following procedure:

$$Q_{cp,i} = q'' \times A_{cp,i} \tag{6}$$

Where, i is the index number for each copper plate. The heat lost by conduction through the wood and to the surrounding by radiation is depicted in Figure 5 and has been estimated using the following equations for each plate.

$$Q_{cond,i} = k_{wood} A_{cp,i} \frac{(T_{s,i} - T_w)}{t} \tag{7}$$

$$Q_{rad,i} = \varepsilon\sigma A_{cp,i} \, (T_{s,i}^4 - T_{surr}^4) \tag{8}$$

The actual heat supplied to each copper plate has been determined by deducting the losses from the total heat supplied to the heater.

$$Q_{actual,i} = Q_{cp,i} - (Q_{cond,i} + Q_{rad,i}) \tag{9}$$

The local convective heat transfer coefficient for each of the copper plate has been calculated using:

$$h_i = \frac{Q_{actual,i}}{A_{cp,i}(T_{s,i} - T_{in})} \tag{10}$$

The average temperature of the heated target surface $T_{s,i}$ has been taken as the average of the readings of the two thermocouples fixed in each copper plate. To calculate h, T_{in} has been considered instead of the bulk temperature or the reference temperature. For a given case (for a given Re, H/d, and orifice-jet plate) T_{in} is fixed. It is measured at the test section inlet, where the air first enters the feed channel. The non-dimensional heat transfer coefficient on the impingement target surface is represented by Nusselt number as follows:

$$Nu_i = \frac{h_i \, d}{k_{air}} \tag{11}$$

The hydraulic diameter has been taken as the diameter of the orifice jet. The data reduction equation for the Nusselt number is considered along with the heat losses by conduction and radiation.

$$Nu_i = \frac{d}{k_{air}} \left(\frac{\dfrac{V^2}{R\,A_{total}} - \dfrac{k_w}{t}(T_{s,i} - T_w) - \varepsilon\sigma(T_{s,i}^4 - T_{Surr}^4)}{(T_{s,i} - T_{in})} \right) \tag{12}$$

4.7 Uncertainty in nusselt number

Temperature of the wood has a very less effect on the uncertainty of heat transfer coefficient due to the large thickness of the wood and also due to the insulation material attached to the wooden block. Temperature of the surroundings and emissivity also has less effect on the uncertainty as the work was carried out in a controlled environment and the temperature of the surroundings was maintained within 21-23 °C through out the experiment. The standard uncertainty in the Nusselt number neglecting the covariance has been calculated using the following equation:

$$\left(U_{c,Nu_i} \right)^2 = \left(\frac{\partial Nu_i}{\partial V} u_V \right)^2 + \left(\frac{\partial Nu_i}{\partial R} u_R \right)^2 + \left(\frac{\partial Nu_i}{\partial T_{s,i}} u_{T_{s,i}} \right)^2$$
$$+ \left(\frac{\partial Nu_i}{\partial T_{in}} u_{T_{in}} \right)^2 + \left(\frac{\partial Nu_i}{\partial A_{total}} u_{A_{total}} \right)^2 + \left(\frac{\partial Nu_i}{\partial d} u_d \right)^2 \tag{13}$$

Uncertainty propagation for the dependent variable in terms of the measured values has been calculated using the Engineering equation Solver (EES) software. The measured variables x_1, x_2 etc. have a random variability that is referred to as its uncertainty. The uncertainty in Nusselt number in the present study has been found to vary between ± 6 % depending upon the jet velocity.

5. Results and discussions

Jet impingement heat transfer is dependent on several flow and geometrical parameters. The jet impingement Nusselt number is presented in a functional form as follows:

$$Nu_i = \left(\frac{h_i d}{k_{air}}\right) = f \left[\begin{array}{c} Re, \ \left(X\!/\!d, H\!/\!d\right), \\ outflow\ orientation \end{array}\right] \tag{14}$$

Where, Re is the flow parameter, jet spacing to the diameter ratio (X/d) is the geometric parameter. The flow exit direction and target surface geometry are also important parameters having a considerable impact on impingement heat transfer.

The X location starts from the supply end of the channel as shown in Figure 7. For the case 1 shown in Figure 10a, flow enters at X/d = 109.3 and exits at X/d = 0. For case 2 (Figure 10b), flow exits at X/d = 109.3. For case 3 (Figure 10c), flow exits at both ends (X/d = 0 and X/d = 109.3). The flow is fully developed before entering the orifice jets. However, in the present study attention is focused on Case – 3 (out- flow passing out in both directions).

5.1 Effect of orifice-jet-plate configuration on feed channel aspect ratio

Figures 12-14 show the local Nusselt number distribution for three orifice-jet plate configurations and for three H/d ratios as a function of non-dimensional location X/d on the heated target surface (for outflow passing in both directions as shown in Figure 10c, and for a given Re= 18800).

Figure 12 shows the effect of feed channel aspect ratio (H/d) on local Nusselt number for Re=18800 for orifice jet plate with centered holes. It can be observed that, H/d=9 gives the maximum heat transfer over the entire length of the target surface as compared to all feed channel aspect ratio studied. H/d=9 gives 1% more heat transfer from the target surface as compared to H/d=5. Whereas H/d=5 gives of 1% increase in heat transfer as compared to H/d=7.

Figure 13 shows the effect of feed channel aspect ratio (H/d) on local Nusselt number for Re=18800 for orifice jet plate with staggered jets. It can be observed that, H/d=9 gives the maximum heat transfer over the entire length of the target surface as compared to all feed channel aspect ratio studied. H/d=9 gives 1% more heat transfer from the target surface as compared to H/d=5, whereas H/d=5 gives of 6% increase in heat transfer as compared to H/d=7.

Figure 14 shows the effect of feed channel aspect ratio (H/d) on local Nusselt number for Re=18800 for orifice jet plate with tangential holes. It can be observed that, H/d=9 gives the maximum heat transfer over the entire length of the target surface as compared to other feed channel aspect ratio studied. H/d=9 gives 3% more heat transfer from the target surface as compared to H/d=7, whereas H/d=7 gives of 6% increase in heat transfer as compared to H/d=5.

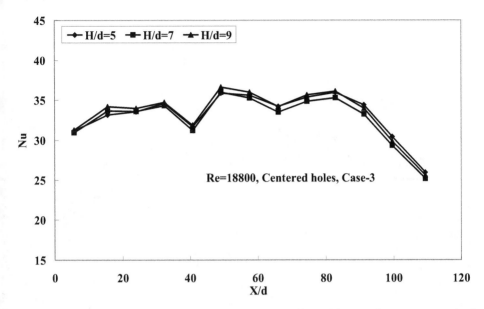

Fig. 12. Nusselt number variation for different aspect ratios and for outflow passing in both directions (for jet-orifice plate with centered holes and for Re =18800)

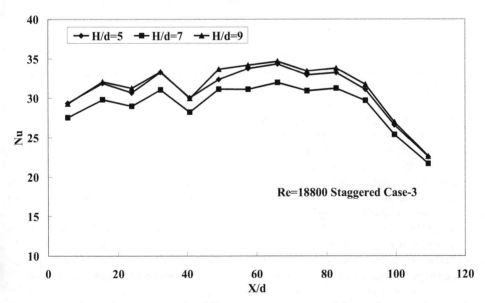

Fig. 13. Nusselt number variation for different aspect ratios and for outflow passing out in both directions (for jet-orifice plate with staggered holes and for Re =18800)

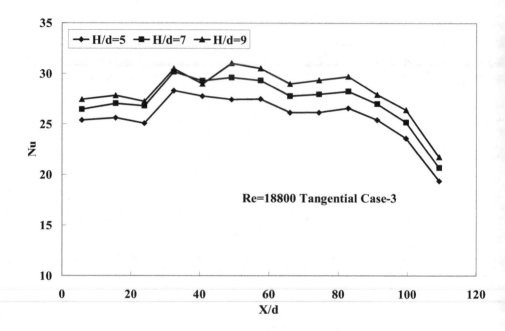

Fig. 14. Nusselt number variation for different aspect ratios and for outflow passing in both directions (for jet-orifice plate with tangential holes and for Re =18800)

5.2 Effect of orifice-jet-plate configuration on local nusselt number

Figures 12-14 also show the effect of the orifice jet plate configurations for different feed channel aspect ratios on local Nusselt number (Nu) along the surface of target surface. Orifice jet plate with centered holes has been found to give better heat transfer characteristics as compared to other plates. For H/d=5, Nu increases in percentage from staggered orifice plate to centered orifice plate by 7% and Nu increases in percentage from tangential orifice plate to staggered orifice plate by 18%. For H/d=7, Nu increases in percentage from staggered orifice plate to centered orifice plate by 11% and Nu increases in percentage from tangential orifice plate to staggered orifice plate by 6%. For H/d=9, Nu increases in percentage from staggered orifice plate to centered orifice plate by 6% and Nu increases in percentage from tangential orifice plate to staggered orifice plate by 10%.

For a given situation (H/d=9, Re=18800 and Case-3) the peak value of local Nusselt number is 36.63 at X/d=49.2 for centered jets. Nu is 34.69 at X/d=66 for staggered jets. Nu is 31.03 at X/d=49.2 for tangential jets.

5.3 Effect of orifice-jet-plate configuration and re on averaged nusselt number

The average Nu is the average of Nu of all 13 copper plates on the target surface for a given situation (i.e. for a given Re, H/d, orifice-jet configuration, outflow orientation). Figure 15 shows the effect of different orifice jet plate configurations on average Nusselt number for outflow orientation Case-3 (outflow passing out in both directions), for different jet Reynolds numbers and for H/d=9. The Nusselt number has been found to increase with increase in Reynolds number. In general, the percentage increase in average Nusselt number in going from Plate-3 to Plate-2 is 11% and in going from Plate-2 to Plate-1 is 11%. This indicates that Plate-1 (centered orifice-jet configuration) gives higher average Nu as compared to other plates.

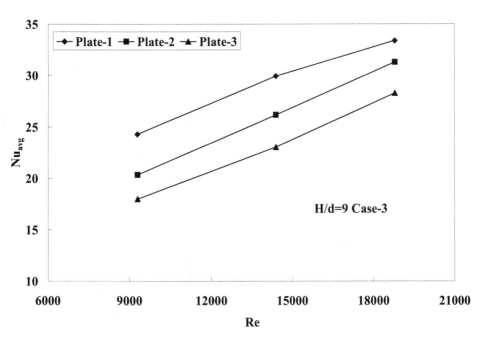

Fig. 15. Average Nusselt number distribution for different jet Re and for different orifice-jet plate configurations (for aspect ratio H/d=9, for outflow passing out in both directions – Case 3)

It is difficult to find out the exact experimental set-up in the literature which has been developed in the present study for comparison of results, however, attempt has been made to make some comparison. Figure 16 compares the results of the present study with archival results of Huang et.al [22] for different jet Re and for different outflow orientations (for a given jet-orifice plate with centered jets). Huang's study focused on multiple array jets, however our study concentrated on single array of centered/staggered/tangential jets (with an inclined target surface). Florschuetz [4] studied experimentally heat transfer distributions for jet array impingement. He considered circular jets of air impinging on heat transfer surface parallel to the jet orifice plate. The air after impingement was constrained in a single direction. Florschuetz presented Nu for centered and staggered hole patterns.

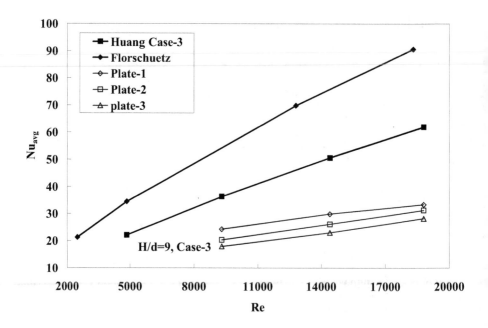

Fig. 16. Comparison of Average Nusselt number of present study with other studies for different jet Re and different orifice-jet plate configurations (for aspect ratio H/d=9, outflow in both directions – Case 3)

6. Conclusions

The above experimental work has discussed in appreciable depth the effect of orifice-jet plate configurations on feed channel aspect ratios (H/d) and on Nusselt number in a channel with inclined target surface cooled by single array of impinging jets (with outflow passing out in both radial directions). In general, it has been observed that Nu is high for higher aspect ratios. For a given plate-1 with single array of equally spaced centered jets and for Re=18800 (outflow passing in both directions), the local Nu for H/d=9 has been found to be greater than Nu of H/d=7 by 5%. The average Nu of plate-1 (centered holes) has been observed to be greater as compared to the Nu of other plate configuration (for a given Re, H/d, and outflow orientation parallel to inlet flow). The averaged Nusselt number has been found to increase with in jet Re regardless of orifice-jet plate configuration. The percentage increase in average Nu has been found to be about 11% with centered holes as compared staggered orifice-jet plate. The percentage increase in average Nu has been found to be about 11% with staggered jet-plate as compared to tangential orifice-jet plate configuration. It can be inferred that from the above results that invariably *(for different combinations impinging jet Re, feed channel aspect ratio, spacing of the target surface from the jet orifices, orifice-jet plate configuration, outflow orientation, etc)* averaged Nu increases with jet impingement cooling. This implies that jet impingement cooling is effective. This eventually results in increase in thermal efficiency and power density of the gas turbines. The observations of the above experimental work offer valuable information for researchers and designers.

7. Acknowledgment

The present work was supported by Research Institute, King Fahd University of Petroleum and Minerals, Dhahran, Saudi Arabia. The authors would like to greatly appreciate the above support. without such support, this work would not have been possible.

8. Nomenclature

$A_{cp,i}$	Area of each copper plate [m²]
A_{total}	Area of all copper plate [m²]
d	Diameter of the orifice jet [m]
h_i	Local convective heat transfer co-efficient [W/m²K]
H	Width of the feed channel [m]
I	Current supplied to heater[Amp]
l	Length of the copper plate [m]
k_{air}	Thermal conductivity of air [W/m.K]
k_{wood}	Thermal conductivity of wood [W/m.K]
Nu_i	Local Nusselt number for each copper plate
Nu_{avg}	Average Nusselt number
q″	Heat flux from the heater [W/m²]
$Q_{cp,i}$	Heat input for each copper plate [W]
Q_{actual}	Actual heat released from target surface [W]
$Q_{cond,I}$	Heat lost due to conduction [W]

$Q_{rad,i}$ Heat lost due to radiation [W]
Q_{total} Total heat input   [W]
Re   Jet Reynolds number
R   Resistance of the heater   [ohm]
t   Thickness of wood block behind the heater   [m]
T_{in} Inlet temperature [°C]
$T_{s,i}$ Surface temperature   [°C]
T_{surr} Temperature of the surroundings   [°C]
T_w Wood block temperature   [°C]
U   Uncertainty
V   Voltage supplied to the heater   [V]
V_{avg} Average velocity of all jets [m/s]
$\forall$ Volume flow rate [m³/s]
X   Distance in the x-direction [m]
θ Inclination Angle [1.5°]

9. Subscripts

cp   Copper plate
i   Index number for each copper plate
j   Jet
w   Wood

10. Greek symbols

ε Emissivity
σ Stefan-Boltzman constant [W/(m²K⁴]
μ Dynamic Viscosity [kg/(ms)]
ρ Density [kg/m³]

11. References

Chupp, P. R. E., Helms, H. E., McFadden, P. W. and Brown, T. R. (1969). Evaluation of internal heat-transfer coefficients for impingement-cooled turbine airfoils. *J. Aircraft*, 6, 203-208.

Florschuetz, L. W., Metzger, D. E., Su, C. C., Isoda, Y. and Tseng, H. H. (1984). Heat transfer characteristics for jet array impingement with initial cross flow. *Journal of Heat Transfer*, 106 (1), 34-41.

Metzger, D. E. and Bunker, R. S. (1990). Local heat transfer in internally cooled turbine airfoil leading edge regions: Part I – Impingement Cooling without Film Coolant Extraction. *Journal of Turbo machinery*, 112 (3), 451-458.

Florschuetz, L. W., Metzger, D. E., Su, C. C., Isoda, Y. and Tseng, H. H. (1981). Stream-wise flow and heat transfer distributions for jet impingement with cross flow. *Journal of Heat Transfer*, 103 (2), 337-342.

Dong, L. L., Leung, C. W. and Cheung, C. S. (2002). Heat transfer characteristics of premixed butane/air flame jet impinging on an inclined flat surface. *Heat and Mass Transfer,* 39 (1), pp. 19-26.

Rasipuram, S. C. and Nasr, K. J. (2004). A numerically-based parametric study of heat transfer off an inclined surface subject to impinging air flow. *International Journal of Heat and Mass Transfer,* 47 (23), 4967-4977.

Beitelmal, A. H., Saad, M. A. and Patel, C. D. (2000). Effect of inclination on the heat transfer between a flat surface and an impinging two-dimensional air jet. *International Journal of Heat and Fluid Flow,* 21 (2), 156-163.

Roy, S. and Patel, P. (2003). Study of heat transfer for a pair of rectangular jets impinging on an inclined surface. *International Journal of Heat and Mass Transfer,* vol. 46, no. 3, pp. 411-425.

Ekkad, S. Huang, Yahoo. and Han, Je-Chin (2000). Impingement heat transfer measurements under an Array of Inclined Jets. *Journal of Thermophysics and Heat Transfer,* 14 (2), 286-288.

Tawfek, A. A. (2002). Heat transfer studies of the oblique impingement of round jets upon a covered surface. *Heat and Mass Transfer,* 38 (6), 467-475.

Seyedein, et. al. (1994). Laminar flow and heat transfer from multiple impinging slot jets with an inclined confinement surface. *International Journal of Heat and Mass Transfer,* 37 (13), 1867-1875.

Yang, Y. and Shyu, C. H. (1998). Numerical study of multiple impinging slot jets with an inclined confinement surface. *Numerical Heat Transfer; Part A: Applications,* 33 (1), 23-37.

Yan, X. and Saniei, N. (1997). Heat transfer from an obliquely impinging circular air jet to a flat plate. *International Journal of Heat and Fluid Flow,* 18 (6), 591-599.

Hwang, J. J., Shih, N. C., Cheng, C, S., et. al. (2003). Jet-spacing effect on impinged heat transfer in a triangular duct with a tangential jet–array. *International Journal of Transfer Phenomena,* 5, 65-74.

Ramirez, C., Murray, D. B., and Fitzpatrick, J. A. (2002). Convective heat transfer of an inclined rectangular plate. *Experimental Heat Transfer,* 15 (1), 1-18.

Stevens, J. and Webb, B. W. (1991). Effect of inclination on local heat transfer under an axisymmetric free liquid Jet. *International Journal of Heat and Mass Transfer,* 34 (4-5), 1227-1236.

Hwang, J. J. and Cheng, C. S.(2001). Impingement cooling in triangular ducts using an array of side-entry wall jets. *International Journal of Heat and Mass Transfer,* 44, 1053-1063.

Hwang, J.J. and Cheng, T. T. (1999). Augmented heat transfer in a triangular duct by using multiple swirling jets. *Journal of Heat Transfer,* 121, 683-690.

Hwang, J. J. and Chang, Y. (2000). Effect of outflow orientation on heat transfer and pressure drop in a triangular duct with an array of tangential jets. *Journal of Heat Transfer,* 122, 669-678.

Hwang, J.J. and Cheng, C. S. (1999). Detailed heat transfer distributions in a triangular duct with an array of tangential jets. *Journal of Flow Visalizationa & Image Processing,* 6, 115-128.

Taylor, B. N. and Kuyatt, C. E. (1994). Guidelines for evaluating and expressing the uncertainty of NIST measurement results. *National Institute of Standards and Technology*, 1297-1303.

Hwang, Y., Ekkad, S. V. and Han, J. (1998). Detailed heat transfer distributions under an array of orthogonal impinging jets. *Journal of Thermophysics and Heat Transfer*, 12 (1), 73-79.

Permissions

The contributors of this book come from diverse backgrounds, making this book a truly international effort. This book will bring forth new frontiers with its revolutionizing research information and detailed analysis of the nascent developments around the world.

We would like to thank Dr. Ernesto Benini, for lending his expertise to make the book truly unique. He has played a crucial role in the development of this book. Without his invaluable contribution this book wouldn't have been possible. He has made vital efforts to compile up to date information on the varied aspects of this subject to make this book a valuable addition to the collection of many professionals and students.

This book was conceptualized with the vision of imparting up-to-date information and advanced data in this field. To ensure the same, a matchless editorial board was set up. Every individual on the board went through rigorous rounds of assessment to prove their worth. After which they invested a large part of their time researching and compiling the most relevant data for our readers. Conferences and sessions were held from time to time between the editorial board and the contributing authors to present the data in the most comprehensible form. The editorial team has worked tirelessly to provide valuable and valid information to help people across the globe.

Every chapter published in this book has been scrutinized by our experts. Their significance has been extensively debated. The topics covered herein carry significant findings which will fuel the growth of the discipline. They may even be implemented as practical applications or may be referred to as a beginning point for another development. Chapters in this book were first published by InTech; hereby published with permission under the Creative Commons Attribution License or equivalent.

The editorial board has been involved in producing this book since its inception. They have spent rigorous hours researching and exploring the diverse topics which have resulted in the successful publishing of this book. They have passed on their knowledge of decades through this book. To expedite this challenging task, the publisher supported the team at every step. A small team of assistant editors was also appointed to further simplify the editing procedure and attain best results for the readers.

Our editorial team has been hand-picked from every corner of the world. Their multi-ethnicity adds dynamic inputs to the discussions which result in innovative outcomes. These outcomes are then further discussed with the researchers and contributors who give their valuable feedback and opinion regarding the same. The feedback is then collaborated with the researches and they are edited in a comprehensive manner to aid the understanding of the subject.

Apart from the editorial board, the designing team has also invested a significant amount of their time in understanding the subject and creating the most relevant covers. They scrutinized every image to scout for the most suitable representation of the subject and create an appropriate cover for the book.

The publishing team has been involved in this book since its early stages. They were actively engaged in every process, be it collecting the data, connecting with the contributors or procuring relevant information. The team has been an ardent support to the editorial, designing and production team. Their endless efforts to recruit the best for this project, has resulted in the accomplishment of this book. They are a veteran in the field of academics and their pool of knowledge is as vast as their experience in printing. Their expertise and guidance has proved useful at every step. Their uncompromising quality standards have made this book an exceptional effort. Their encouragement from time to time has been an inspiration for everyone.

The publisher and the editorial board hope that this book will prove to be a valuable piece of knowledge for researchers, students, practitioners and scholars across the globe.

List of Contributors

Konstantinos G. Kyprianidis
Chalmers University of Technology, Sweden

Marek Dzida
Gdansk University of Technology, Poland

Roberto Biollo and Ernesto Benini
University of Padova, Italy

Marco Antônio Rosa do Nascimento and Eraldo Cruz dos Santos
Federal University of Itajubá – UNIFEI, Brazil

Liqiang Duan, Xiaoyuan Zhang and Yongping Yang
School of Energy, Power and Mechanical Engineering, Beijing Key Lab of Energy Safety and Clean Utilization, Key Laboratory of Condition Monitoring and Control for Power, Plant Equipment of Ministry of Education, North China Electric Power University, Beijing, People Republic of China

Mario L. Ferrari and Matteo Pascenti
University of Genoa – Thermochemical Power Group (TPG), Italy

Ene Barbu
National Research and Development Institute for Gas Turbines COMOTI, Romania

Valeriu Vilag, Jeni Popescu, Silviu Ionescu, Adina Ionescu, Romulus Petcu, Cleopatra Cuciumita, Mihaiella Cretu, Constantin Vilcu and Tudor Prisecaru
National Research and Development Institute for Gas Turbines Comoti, Romania

Li Pan
Queen's University Belfast, U.K.

Diango A., Périlhon C., Danho E. and Descombes G.
Laboratoire de génie des procédés pour l'environnement, l'énergie et la santé Conservatoire national des arts et métiers, Paris, France

Luai M. Al-Hadhrami, S.M. Shaahid and Ali A. Al-Mubarak
Associate Professor Center for Engineering Research, Research Institute, King Fahd University of Petroleum and Minerals, Saudi Arabia

Printed in the USA
CPSIA information can be obtained
at www.ICGtesting.com
JSHW011429221024
72173JS00004B/724